Math and Students With Learning Disabilities

Math and Students With Learning Disabilities

A Practical Guide To Course Substitutions

Paul Nolting, Ph.D.

Academic Success Press
Bradenton, Florida 34206

Math and Students With Learning Disabilities: A Practical Guide to Course Substitutions

BY PAUL D. NOLTING

Published by Academic Success Press, Inc.

Printed in the United States of America

First Printing

First Edition

ISBN: 0-940287-24-2

Table of Contents

Figures:

Preface

Students with learning disabilities (LD) attending colleges and universitie continues to increase throughout the United States and Canada. Presently there is a national movement towards increasing mathematics course requirements for certain majors or for graduation. Mathematics is already one of the most difficult college subjects for any student and can be extremely difficult for students with LD. Some colleges require students to pass up to five non-credit and credit mathematics courses to graduate.

Many students with LD have received appropriate learning and testing accommodations for mathematics courses. Some of these accommodations include extended test time, note-takers, tape recorders, enlarged text, calculators, fact sheets and private test areas. However, even with appropriate accommodations and substantial effort some students with LD cannot pass their required mathematics courses.

This book is written for counselors, mathematics instructors and administrators to provide information on mathematics course substitutions. The focus of this book is on answering the following three questions:

- Which students with LD can pass their required mathematics courses
- Which students with LD should be granted course substitutions
- Which courses are appropriate substitutions

Different types of learning disabilities are explained including their effects on mathematics learning. Reasons for students with LD being unable to pass mathematics even with appropriate accommodations are discussed.

Suggested course substitution procedures are detailed and a check list for Disabled Student Service Providers is presented. The importance of this information is to give students the opportunity to graduate when their uncompleted mathematics course(s) are not an essential part of their program.

A review of actual student course substitution requests from different college are discussed in case histories. These eight students with LD case histories range from simple to complex. A learning disability profile is presented for each student along with the accommodations used. The outcome of each case is explained and may be generalized to other learning disabled students.

Helping students with LD become successful in mathematics is the responsibility of administrators, counselors, and mathematics instructors. This group needs to work together to provide adequate financial resources, counseling for students with disabilities, and appropriate accommodations. When appropriate accommodations are not successful students should have the opportunity to apply for a mathematics course substitution. Information is this book can help institutions make the decisions about granting a mathematics course substitution.

Chapter 1

Relating Learning Disability Definitions To Course Substitutions

1

The number of students with learning disabilities (LD) at community colleges and universities are increasing throughout the United States and Canada. Improved special education programs in elementary schools, middle schools, and high schools have helped students with LD graduate. White, Alley, Deshler, Schumaker, Warner, and Clark (1982) estimate that about 67 percent of high school students with LD plan to attend college. In 1982, about six percent of the college freshman class reported learning disabilities (Beirne-Smith & Decker, 1989). Dale and Schmidt (1987) reported that more than 500 two- and four-year institutions provided services to students with LD. Jarrow (1987), director of the Association on Higher Education and Disability, has stated that students with LD are "the single largest contingent of students with disabilities being served on American campuses" (p.46).

Learning mathematics is considered extremely difficult for many students with LD (Hughes and Smith, 1990) and is sometimes impossible at ceratin course levels (Nolting, 1991). Vogel (1985a) noted in her survey of 32 college students with LD that 62 percent reported learning difficulties with word problems and geometry. She also reported that about half of the students had difficulty with multiplication tables. Bireley (1986) reported that 25 percent of his college sample had problems with basic math skills, while 67 percent had difficulty in higher math. The higher math courses were not defined but can be assumed to include algebra, geometry, and statistics.

Skrtic (1980) reported that 7th- and 8th-grade students with LD performed worse than their peers without LD on the quantitative concepts and applications problems on the **Woodcock-Johnson Psychoeducational Battery** (Woodcock &Johnson, 1977). The scores of the students with LD were equivalent to fourth graders. Their fourth grade performance level indicated that these students had not progressed to the higher stage of problem solving. Another comparison study of 10th grade students with and without LD on mathematics achievement test scores indicated that students with LD did better on problems requiring literal use of numbers and poorer on problems requiring mathematical concepts. The students with LD had a lower degree on mastery than nondisabled students (Algozzine, et. al, 1987).

Montague, Bos, and Doucette (1991) sum up their literature findings by stating that these students are different from their nondisabled peers, not only in general mathematics ability, but also in affective, cognitive, and metacognitive characteristics.

Zentall (1990) adds to these finding by stating that students with LD did worse than their nondisabled peers on concepts/application problems. Hutchinson (1993), in her review of the literature states,"... LD students frequently lack essential general problem-solving and domain-specific knowledge, exhibit deficits in executing specific mathematical strategies, and fail to use self-regulation." The performance of students with LD on the **Wide Range Achievement Test** arithmetic sections decreased over time during ages eight to 16 (Ackerman, et. al., 1986a). Research now indicates that difficulty with mathematics due to LD emerge early in education and worsens over time (Zentall & Ferkis, 1993, & Mercer, 1991). This decline in achievement occurs even when students with LD are spending up to one-third of their time in resource rooms used for mathematics remediation and instruction (Carpenter, 1985). Many times, after having difficulty passing basic algebra, these students take consumer math, business math, or other courses based on arithmetic skills to graduate from high school. These students attend colleges and universities where their mathematics learning problems continue to persist. However, many of these students were successful in all their other college courses.

The characteristics of postsecondary students with LD are similar to their counterparts in elementary and secondary schools. Some of these students were diagnosed as LD early in their school life, but many of these students were diagnosed as LD while in college. Zawaiza and Geber (1993), in their article on math students with LD in community colleges some up their characteristics by stating:

> *... like younger students with learning disabilities, these students still tend to be slower and less competent as learners compared with their achieving peers. This is especially the case when LD students have to apply basic language mathematics skills while pursuing occupational or liberal arts studies.*
>
> *Relatively little empirical data exist on these students' mathematical performance. Accordingly, having obtain opportunities for higher education, these students confront significant barriers to successfully pursuing postsecondary academic or vocational goals. Until we understand their difficulties more fully and are able to*

develop suitable interventions, effort to assist these students will continue to rely on inefficient trail-and-error approaches.

Passing mathematics courses is a graduation requirement for most colleges and universities. However, students with LD, who fail mathematics, may not graduate even though mathematics is not an essential part of their degree. The research clearly indicates that students with LD have difficulty learning mathematics. In contrast, many students with LD do complete their mathematics courses and graduate. The challenge is identifying students with LD who cannot pass mathematics among those students with LD who can!

Colleges and universes are granting mathematics course substitutions to students with LD. These students are allowed to graduate and work in their field of study. There is limited information in the literature on the process, procedures and rational for granting mathematics course substitutions. Brinckerhoff et al., (1993) indicates that policies for course subsititutions are typiclly not in place and rigid course requirements can also deny access to other groups of students with disabilities.

Postsecondary institution administrators are struggling with three decisions:

- which students with LD can pass their required mathematics courses
- which students with LD should be granted course substitutions
- which courses are appropriate substitutions.

The information in this book can be used as a guide to answering these three questions.

Learning Disability Definitions

In describing learning disabilities, there is a clear distinction between a conceptual definition and an operational definition of learning disabilities. Conceptual definitions focus on the theoretical aspects of a learning disability and should clearly state the exact aspects of a learning disability. Operational definitions are based on conceptual definitions and can be used to qualify students as learning disabled. Operational definitions usually have a statistical requirement that the student's test scores must meet in order to qualify the student as learning disabled. Operational definitions can vary widely from college to college and state to state. Some of the conceptual definitions described below to provide the reader with a better understanding of learning disabilities.

The term learning disability (LD) is usually used to describe a broad range of neurological dysfunctions. Since learning disabilities are invisible, this term is often misunderstood. The Federal Register defines a LD as:

> *A disorder in one or more of the basic psychological processes involved in understanding and use of language, spoken or written, which may manifest itself in an imperfect ability to listen, think, speak, read, write, spell, or to do mathematical calculations. The term includes such conditions as perceptual handicaps, brain injury, minimal brain dysfunction, dyslexia, and developmental aphasia. The term does not include individuals who have learning problems which are primarily the results of visual, learning or motor handicaps, or mental retardation, or environmental, cultural, or economic disadvantage.*

A specific learning disability, according to rule 6H-1.041 for the Florida community college system, is:

> *A disorder in one or more of the basic psychological or neurological processes involved in understanding or in using spoken or written language. Disorders may be manifested in listening, thinking, reading, writing, spelling, or performing arithmetic calculations. Examples include dyslexia, dysgraphia, dyscalculia, and other specific learning disabilities in the basic psychological or neurological process. Such disorders do not include learning problems which are primarily due to visual, hearing, or motor handicaps, to mental retardation, to emotional disturbance, or environmental deprivation.*

Hammill (1990) stated that the most widely accepted definitions of learning disabilities was developed by the National Joint Committee on Learning Disabilities (1988):

> *Learning disabilities is a general term that refers to a heterogeneous group of disorders manifested by significant difficulties in the acquisition and use of listening, speaking, reading, writing, reasoning, or mathematical abilities. These disorders are intrinsic to the individual,*

presumed to be due to central nervous system dysfunction, and may occur concomitantly with other handicapping conditions (for example, sensory impairment, mental retardation, serious emotional disturbance) or with extrinsic influences(such as cultural differences, insufficient or inappropriate instructions), they are not the results of those conditions or influences.

The definition of learning disabilities suggests that there is a neurological aspect causing the learning disability. There has been debate in the literature on this subject for a number of years. Bigler (1992), in his article, **The Neurobiology and Neuropsychology of Adults Learning Disorders,** states that magnetic resonance imaging (MRI) and eletrophysiological tests found anatomic irregularities in adults with LD compared to controls. Adults with LD have a variety of underlying neurobiologic irregularities and/or abnormalities that are permanent. Bigler (1992) conclude his article by stating, "Adults with learning disabilities have a neurologically based disorder." His conclusions are important; he proves the existence of LD and that in adults it is a permanent condition. This means that adults with LD can not be "fixed;" their learning problems are real. On the bases of this information granting appropriate mathematics course substitutions is justified.

Some colleges and universities have their own operational definitions of learning disabilities. Disability Support Services coordinators can use these definitions to determine if a student has a learning disability and is eligible to apply for a mathematics course substitution. If a college or university does not have an operational definition of a learning disability, then the coordinator can use outside documentation to determine if the student is disabled. This outside documentation could be high school reports or reports from outside evaluators. In some cases the college or university learning specialist will assess students for learning disabilities using a prescribed protocol. Once students are determined to have a learning disability and demonstrate difficulty in learning mathematics, then they can request a mathematics substitution.

In general, a student with a learning disability effecting mathematics learning has average to above average intelligence and has difficulty in one or more of the basic neurological functions. Neurological dysfunctions could be in the form of verbal reasoning, long-term memory, long-term retrieval, short-term memory, processing speed, auditory processing, visual processing, or fluid (abstract) reasoning. These disorders may impair mathematics learning (dyscalculia) along with reading (dyslexia), writing (dysgraphia), thinking, and spelling. Even if the student is not diagnosed with dyscalculia, these neurological dysfunctions interfere with mathematics learning and the student's ability to demonstrate knowledge on tests. In some cases there is a difference between the student's mathematics aptitude score and the mathematics achievement score. However, the adult learner with excellent compensation skills, such as diligent studying with constant tutoring, may not show any difference between their mathematics aptitude and mathematics achievement. In fact, some adult students will have a positive aptitude - achievement difference showing that their achievement is 1.5 to 2.0 standard deviations (SD) above their aptitude score. These students have maximized their mathematics achievement and probably cannot learn any more mathematics.

Learning disabilities do not include learning problems primarily due to physical disabilities, emotional disturbance, or lack of previous opportunity for learning. Learning disabilities cannot be "cured". But in many cases, learning disabilities can be circumvented through learning and testing accommodations.

Types of Mathematics Learning Disabilities

Students with LD having difficulty learning mathematics generally fall into three areas:

- Difficulty doing the actual calculations
- Difficulty understanding the mathematical concepts
- Difficulty doing calculations and understanding mathematical concepts

Some students have good calculation skills for problems that require limited abstract reasoning and may be improved with practice. Zentall and Ferkis (1993), in their review of the research, suggest that cognitive ability may not be a major part of computational performance. However, cognitive style factors, such as attention and visual processing speed, have been associated with poor computational abilities. Other students may have poor calculations skills but understand abstract mathematics concepts. Their poor calculation skills interfere with their ability to apply the mathematical concepts to solving problems. This is similar to an English student who has excellent conceptual writing skills but has difficulty spelling words. The final area consists of students who have poor calculation skills and conceptional skills but may have a 3.00 GPA, as they major in English.

In most cases, the student with LD will have average calculations skills, especially in arithmetic; but they have difficulty understanding algebraic concepts. These students usually start in developmental mathematics courses or the lowest level algebra course and try to work their way through the algebra course sequence. Somewhere along the course sequence, they repeatedly fail a course that is either a prerequisite to other mathematics courses or the mathematics course needed for graduation.

Other characteristics of students with LD who have difficulty learning mathematics are:

- Problems in a series of mathematical steps to solve a problem
- Inability to apply mathematics concepts to word problems
- Difficulty solving oral problems
- Difficulty visually reading graphs
- Difficulty understanding the instructor
- Inability to take notes and understand the lecture at the same time
- Poor mathematics study skills
- High test anxiety

Chapter 2

Reasons for Mathematics Learning Problems

2

Students with learning disabilities who have severe mathematics learning difficulties may have one or more major neurological dysfunctions that effect information processing. These information processing disorders block their ability to obtain valuable information in learning mathematics and/or in demonstrating their mathematics knowledge on tests. Most of these students with LD did not have a severe difference between their general or mathematics aptitude/IQ and mathematics achievement scores. However, even with learning and testing accommodations, these students cannot pass their mathematics courses.

The Disabity Support Service Provider must review the student's test records to locate the processing disorder(s) that effects learning. The recommended test records to review inclulde an IQ test, such as the **Wechsler Adult Intelligence Scale - Revised** (WAIS-R, 1981), and a psycho-educational test battery, such as the **Woodcock-Johnson Psycho-Educational Battery - Revised** (WJPB-R, 1989). The **Detroit Test of Learning Aptitude - Adult** (1992), can also be helpful in assessing processing difficulties. If the student does not have the test results to indicate processing disorders, then additional tests need to be administrated. Even if your college or university does not offer assessments for learning disabilities, the testing specialist or disablity support service counselor may wish to assess these students to better understand their processing deficits.

The results from the **WAIS-R** can give you some insight about the student's processing disorders and suggest if the student will be successful in college studies. The **WAIS-R** has a Full Scale IQ that is composed of a Verbal IQ and a Performance IQ. The Verbal IQ is broken down into six subtests:

- Information
- Digit Span
- Vocabulary
- Arithmetic
- Comprehension
- Similarities

The Performance IQ is determined by performance on five subtests:

- Picture Completion
- Picture Arrangement
- Block Design
- Object Assembly
- Digit Symbol

The arithmetic, coding, information, and digit span subtests scores correlate significantly with the **Wide Range Achievement Test** arithmetic subtest scores (Ackerman et al., 1986a).

Vogel (1991), in her study of college students' **WAIS-R** scores, discovered that, in general, **WAIS-R** scores were not predictors of success for her group. However, the Comprehension and Similarities subtest scores were correlated to GPA and are probably indicators of academic success. The Comprehension subtest scores measure social judgment and common sense reasoning. The Similarities subtest scores measure verbal abstract reasoning, categories and relationships. A Verbal IQ greater than a Performance IQ may indicate potential for success. Evidence of a learning disability may be most apparent in Digit Span and Arithmetic scores that are lower than other subtest scores. The Digit Span subtest measures short-term auditory memory and concentration. The Arithmetic subtest score measures attention, concentration, and numerical reasoning. A large discrepancy between Verbal and Performance scores on the **WAIS-R** may not indicate a learning disability.

Students who score low on Arithmetic, Picture Arrangement, and Block Design subtests relative to other subtests may have severe problems learning algebra. Picture Arrangement measures hypothesis testing and sequential logical thinking. Block Design measures discrimination, perceptual organization, concrete and abstract visual problem solving. Low scores on one or more of these subtests indicate abstract reasoning processing difficulties that effect the students ability to understand mathematics concepts. This does not mean that all students with average Arithmetic, Picture Arrangement, or Block Design subtest scores will be successful in algebra. Remember: *These subtests are not perfect measurements of mathematical ability.*

The **WJPB-R** (1989) is a good psycho-educational battery to assess processing disorders that cause mathematics learning problems with adult students with LD. The **WJPB-R** measures the following processing areas:

- Long-Term Retrieval

- Short-term Memory
- Processing Speed
- Auditory Processing
- Visual Processing
- Comprehension-Knowledge
- Fluid Reasoning

In a previous text, **Math and the Learning Disabled Student: A Practical Guide for Accommodations** (Nolting, 1991) there is a discussion of how different processing disorders effect mathematics learning. His book also suggests appropriate learning and testing accommodations for different types of processing disorders. An explanation of each processing disorders may help the reader understand the severity of it's effect on mathematics learning.

Processing Disorders

Visual Processing Speed/Visual Processing

Learning disabled students who have visual processing speed disorders will have difficulty learning mathematics. Batchelor et al. (1990) reported that visual tracking, which is similar to visual processing speed, is a very important factor in mathematics achievement. In this case visual processing speed refers to the speed of working with understood mathematics symbols and numbers. In other words, this is the speed at which students can copy down recognizable numbers and symbols.

A student's visual processing speed will effect how fast he/she can copy notes from the board and take a mathematics test. Kirby and Becker(1988) discovered that students having mathematics learning problems were slow in doing their mathematics problems. These students had slow visual processing speed. Many of these students performed accurate mathematics computations but could not do them quickly.

Learning disabled students who have visual processing disorders will have trouble quickly identifying symbols. Rourke and Strange (1983) and Strange and Rourke (1985) identified neuropsychologically distinct subtypes of this arithmetic dysfunction. These students had normal verbal and auditory-perceptual skills but demonstrated deficiencies in both visual-spacial and tactile-perceptual skills. In other words, they had problems with the ability to recognize and remember, in sequence, complex mathematics symbols and numbers that may not be known. These problems also will cause difficulty in reading tests and with the mathematics textbook. Mistakes can occur, when students miscopy notes and misread the textbook or test questions.

Short-Term Memory/Auditory Processing

Students with learning disabilities who have short-term memory/ auditory processing difficulties may also have difficulty learning mathematics. This problem becomes apparent when listening to a mathematics instructor explain the steps of working a problem. For example: Students may forget the mathematics problem steps before writing them in their notes, or may recall and write down the mathematics problem steps in the wrong order. Either mistake will cause difficulty in understanding the mathematics problem and using class notes as a homework guide.

Another problem these students have is retaining mathematics concepts long enough in short-term memory to understand its application. If a series of related mathematics concepts are discussed in the mathematics textbook, these students will have difficulty remembering one concept long enough to apply it to the next concept.

Students with poor short-term memory will most likely have difficulty with mathematics word (story) problems. These students will spend more time re-reading the word problems to understand the question. After understanding the question, they will spend additional time setting up the equation to solve the problem. In general, students with poor short-term memory have difficulty in most of their reading areas. According to Blalock (1980), about one-third of young adults with learning disabilities have some type of auditory memory problem.

Students with auditory processing difficulties have problems telling the difference between certain sounds of words. Auditory processing is a measurement of a student's ability to put together different sound patterns into words and evaluate the difference between auditory patterns. Adult learning disabled students may continue to have difficulty understanding tasks or concepts that require good auditory processing skills. Meyers (1987) explains that not only can reading efficiency be affected by an auditory processing deficit, but the student's ability to understand lectures and oral directions is dramatically affected as well. In fact, the more severe the auditory processing deficits, the more misunderstanding of the lecture material. These students will "miss" some of the words in a lecture or replace words with incorrect words. When this occurs, the student will have difficulty understanding the instructor and writing his/her notes. Meyers (1987) explains that these students will have major difficulty understanding lectures in distracting situations. Auditory processing problems sometimes decreases the ability to read the mathematics textbook or listening to lectures which can cause the student to misunderstand mathematics concepts.

Gifted Students With Learning Disabilities

To better understand how processing disorders effect learning, Waldron and Saphire (1989) compared the intellectual patterns of gifted learning disabled students to gifted non-learning disabled students. The gifted learning disabled students performed significantly poorer in perceptual areas, such as visual and auditory discrimination, visual and auditory sequencing, short-term auditory memory, and visual-spacial skills, compared to the non-learning disabled gifted students. No difference existed between the groups in visual memory skills or listening comprehension. The gifted learning disabled students also had comparative weakness in mathematics, reading, and spelling.

Daniels (1983) examined the strengths of gifted learning disabled students using the **WAIS-R** and discovered that their highest scores were on the Similarities test. The Similarities test measures reasoning and conceptual thinking ability. Their lowest scores were generally on the Arithmetic and Digit Span tests. The Low Arithmetic test scores may indicate poor mathematics skills/or reasoning and an inability to concentrate or listen. The Low Digit Span test scores may indicate poor short-term memory, poor attention or lack of concentration.

The research on gifted learning disabled students supports the concept that even extremely intelligent students can have difficulty learning mathematics. This can give us insight into the learning problems of students who make good grades in everything but mathematics. The visual discrimination, memory problems, and arithmetic problems exhibited by these students interfere in the learning of mathematics.

> *Gifted or non gifted students may "live off" certain aspects of their IQ (learning ability) to achieve a certain level of mathematics competency. However, these students may eventually take a mathematics course where their IQ (learning ability), even with accommodations, will not have enough "power" to compensate for their other processing deficit problems.*

At the same time these students are excelling in their other courses and college activities but can't pass the required mathematics courses.

Mathematics instructors usually start questioning a student's effort and motivation in learning mathematics without understanding the student's learning disability. To help these students and the instructors, we must understand to what extent processing deficits effect each student's mathematics learning and know when students have maximized their mathematics learning.

In general, processing disorders hinder students in receiving lecture information and taking tests. The lecture information could be learned in the incorrect order or totally forgotten before the information is recorded as notes. Processing difficulties especially, when taking a mathematics course, can cause problems learning a sequence of mathematics concepts.

Students may have above average grades in all their courses, but mathematics presents a special concern. These students can make a "D" or "F" in mathematics courses but make an "A" or "B" in their other courses. Many of these students may have a learning disability along with poor mathematics study skills and test anxiety. The fact that some of these students are considered gifted in specialty areas, such as humanities, art, English and athletics, further confuses the problem.

Severe Processing Disorders

Long-Term Retrieval

Long-term retrieval problems can severely impact on mathematics learning. Undergraduates with good longer memory spans reported better delayed recall of relevant information in algebra story problems than students with LD. The students with LD had less success with story problems than non-LD students (Cooney and Swanson, 1990). Zawaiza & Gerber (1993) in their study of community college students with LD supports the claim that students who have difficulty with their working memory (Long-Term Retrieval) have difficulty solving mathematics problems.

Long-term retrieval is not to be confused with the amount of information that is available which is called comprehension-knowledge. Long-term retrieval can be measured over a period of minutes or up to seven days. The Long-Term Retrieval cognitive cluster score of the **WJPB-R** measures the effects of recalling information after learning has been consistently interrupted over a period of several minutes. The Delayed Recall-Memory for Names and Delayed Recall-Visual-Auditory Learning subtest measures long-term retrieval from one to seven days. The **WJPB-R** gives separate scores for Long-Term Retrieval after a few minuets and after several days. The **WAIS-R** does not measure long-term retrieval but does have subtests that measure long-term memory.

Mather (1991) suggests that when assessing Long-Term Retrieval the intervening time is not the essence, but how much other tasks have effected working memory during the test and the ability to recall that information. Long-term retrieval tests, compared to the short-term memory test, require students to store and retrieve pictures and symbols over a period of time. Short-term memory tests require students to recall information immediately after it is given without any interruptions. Mather indicates that students with long-term retrieval problems have difficulty with paired-association, tasks such as remembering the letters of the alphabet or their multiplication tables. This also suggests future difficultly in reading and mathematics.

Learning disabled students with long-term retrieval problems may listen to a mathematics lecture and understand each step as it is explained. However, when the instructor goes back to a previous step discussed several minutes ago and asks the LD student a question, the student cannot explain the reasons for the step("... but I knew it a few minutes ago!"). These students have difficulty remembering series steps long enough to understand the concept. Students with short-term memory problems forget the step as soon as the instructor explains it.

Another problem occures when students are inconsistent in learning new facts or concepts. Students may be able to learn how to work fractions one day but have difficulty using or recalling this process several days later. However, when they are taught fractions again, it is easier for them to learn the steps. These students need constant reviewing of mathematical concepts during homework or tutoring sessions just to remember how to work the previous problems.

Long-Term Memory

Long-term memory is a component of mathematics learning in that it acts as an information source for mathematics formulas, mathematics vocabulary, the recognition of different problem types, and algorithms. Deficits in long-term memory result in poor arithmetic achievement (Dinnel, Glover, Ronning, 1984).

The same memory loss can happen working multiplication tables. Students may forget the multiplication table, but demonstrate the concept of multiplication. Students may have poor achievement in mathematics calculations but may have average or above average mathematics reasoning ability. Often the result of a poor long-term memory is that while at one time the student understood how to work a problem, he/she forgot the concept or mechanical calculation needed to solve the problem on demand.

The Information subtest of the **WAIS-R** measures long-term memory and general factual knowledge. A low score on the information subtest indicates a limited fund of general knowledge, remote or long-term memory, or the ability to access that memory. Limited educational or cultural experiences may also cause low scores. I have worked with students with LD who had significantly low Information subtest scores who had tremendous difficulty remembering how to work algebra problems. They could not retain the algebra concepts over a period of time, and the concept had to be constantly reviewed before learning new mathematics material.

The **WJPB-R's** Comprehension-Knowledge tests of Picture Vocabulary and Oral Vocabulary measure stored knowledge including metacognition, communication, and different types of reasoning based on previously learned procedures. It has been called crystallized intelligence and long-term memory. Students with LD who score low on this test usually have poor vocabularies and sometimes limited background knowledge (Mather, 1991).

Most of the research in this area is related to reading comprehension instead of mathematics. However, the reading research points to students who have a limited knowledge base experiency difficulty in reading comprehension. This long-term memory problem can also effect mathematics students who need a memory base in basic mathematics skills and concepts to solve problems. Without this knowledge base, these students cannot remember necessary vocabulary words to comprehend reading materials. Nor can mathematic students totally remember alogrithems or concepts to solve algebraic equations.

Students with LD having this problem may know mathematics properties one day and several weeks or months later do not understand how to work the same problems. These student might do fine on mathematics quizzes but fail their major mathematics tests. Since learning mathematics is cumulative, these student usually do worse on each successive mathematics test. However, in some of their other subjects areas, which do not have cumulative tests, these same students might be making an "A" and "B".

Fluid/Abstract Reasoning

Another type of learning disability is manifested in problems with fluid reasoning or thinking. These problems usually occur when abstract reasoning is required to move some mathematics concept from its property or principle to its application. Fluid Reasoning is the capability to reason in a novel situation, using inductive, deductive, conjunctive, disjunctive and other forms of reasoning (Woodcock, 1990). The **WJPB-R** Fluid Reasoning cluster measures a student's ability to figure out the missing parts of a mathematics-related logic puzzle and the student's ability to recognize, use, and explain rules to logic puzzles. In short, the Fluid Reasoning cluster measures nonverbal abstract reasoning, conceptual thinking, and problem solving skills. For example: Students may have difficulty understanding and applying a formula to a new homework problem. Students may remember the concept but forget how to use it in a problem. In general, students may have difficulty understanding abstract mathematics formulas and generalizing the formula's use to their homework or test problems.

Hessler (1984) indicates that the performance on the reasoning score (fluid reasoning) is very closely related to mathematics achievement. Students who score high on the Fluid Reasoning cluster have good mathematics achievement scores. Students who score low on the Fluid Reasoning cluster will, in most cases, do poorly on mathematics achievement. In addition, students who score low in the Fluid Reasoning cluster frequently will have difficulty solving problems that require conceptualization and abstract mental processing.

Bley and Thornton (1981) suggest that students with learning disabilities in fluid reasoning may have more trouble with mathematics than any other area. Learning disabled students with reasoning problems often have difficulty verbalizing what has been learned, relating the concepts with symbolic language, and auditorially or visually understanding the instructor's explanation. Difficulties with abstract reasoning make it very difficult to sequence the steps of a problem logically (Thornton & Bley, 1982). Students may understand the first concept of solving a problem but will have difficulty with steps using multiple associations.

Fluid reasoning is an excellent predictor of student success in algebra. Teaching students how to improve their fluid reasoning is very difficult (Mather, 1991). Students with severe fluid reasoning deficits can only go so far in their algebra courses, because it becomes impossible to learn any more algebra. It is extremely difficult to provide appropriate learning and testing accommodations for these students since the processing disorder is an internal organic disfunction. This processing disorder is not in receiving the information to be analyzed but it is the actual cognitive analyzation of mathematical concepts.

Linear Learning

Usually students with LD who take mathematics classes experience more difficulty learning mathematics than other subjects. This is especially true of students with severe processing deficits. This is primarly because mathematics requires linear learning. This means that material learned one day is used the next day and next month. The severe processing deficits cause a major interruption in linear learning. These same students in many nonlinear courses(in which material learned one day can be forgotten after the test) will not have as much difficulty learning the material.

For example: In some non-linear courses — such as social sciences — misunderstanding the first part of a lecture doesn't mean a student will not understand the rest of the class material. However, in a mathematics class, if a student misunderstands the first concept taught, he/she is probably lost for the remainder of the class. In a non-linear class, a student can fail the first major test but may not need to understand that material for the next test. In a mathematics class, if a student fails the first test, then he/she must learn that material to pass the next test. If the student has trouble with long-term retrieval, long-term memory or fluid reasoning, there are major problems with being re-tested on previous material. The student has to go back and relearn that material, generalize, then learn new material to pass the next test.

This discussion on the reasons students with LD have difficulty learning mathematics has focused on specific processing disabilities. Some students with LD have a combination of severe and less severe processing disorders. Disablility Support Service Providers must be able to understand and explain to faculty and administrators how each processing disorder effects mathematics learning. They must know which students have severe processing disorders in order to start preparing a case for a potential course substitution. They also must understand the extent other affective factors, such as test anxiety, motivation, and mathematics study skills, impact on mathematics learning.

Chapter 3

Using Learning and Testing Accommodations to Substantiate Course Substitution Needs

3

When a math course substitution is requested, the Disability Support Service Provider must be able to substantiate that the student with LD has already tried the courses(s) with appropriate learning and testing accommodations. The service provider then can state that even with appropriate accommodations and student effort the student has reached his/her highest mathematics achievement level. Continuing to have the student fail in his/her attempts to master mathematics will be psychologically damaging and will not improve his/her mathematics knowledge. Demonstrating that the student has used the suggested accommodations, studied diligently, used tutor services, and sought out the instructors help will verify that everything possible has been done to help the student. If you cannot prove these facts, then the substitution committee most likely will have the student repeat the mathematics course by "trying harder" with better accommodations. Understanding these factors makes it imperative that the Disability Support Service Provider allow students the maximum appropriate accommodations and tutorial services available. Remember, the type of learning and testing accommodations must be documented to demonstrate that the student had the best opportunity to learn and prove the subject knowledge on tests.

Learning accommodations should allow students with LD the same access to course material as non-disabled students. Course materials can be in the form of lectures, labs, field trips, library resources, and books. Learning accommodations may be the same or different for various types of students with LD.

The extent of learning and testing accommodations given by institutions will vary. Some institutions' philosophy is to give extensive learning and testing accommodations to the point of changing the course curriculum, while the philosophy of other institutions is to provide adequate accommodations and to substitute the course instead of changing the course curriculum. A brief review of learning and testing accommodations is presented. A more extensive explanation of learning and testing accommodations is discussed in **Math and the Learning Disabled Student: A Practical Guide to Accommodations** (Nolting, 1991).

Learning Accommodations Based on Processing Disorders

Visual Processing Speed/Visual Processing

Dr. Hessler (1984), author of **Use and Interpretation of the Woodcock-Johnson Psycho-Educational Battery**, suggests that visual processing speed is more highly related to mathematics achievement than other measures of perception. Many students with LD are processing material at the one-to-ten percentile level. This means that 90 to 99 percent of the other students visually process material faster or better than students with LD. For students with visual processing or processing speed disorders, taking mathematics notes is a major problem. They usually take very few notes or conversely, try to write everything down. They can have number reversals or transpose parts of an equation while writing notes. They may also have difficulty keeping problem steps properly aligned in the correct columns.

Calculators can be another learning accommodation for students with LD. Calculators can reduce the problems of misreading numbers and keeping decimals places straight. Calculators can solve the speed and accuracy problems that are prevalent with visual processing speed and visual processing problems. This will allow the student to have a better focus on mathematics concepts, instead of worrying about careless errors. It is suggested that students use calculators with large LED read outs and in severe cases, talking calculators.

Three lecture accommodations for learning disabled students with visual processing speed or visual processing problems are note takers, tape recorders, and handouts. Students with LD may need only one or all of these learning accommodations. Students with LD may only need these accommodations in a mathematics class and not in their other subjects.

Short-Term Memory/Auditory Processing

Students with LD having short-term memory/auditory processing problems will have difficulty learning mathematics. Students who have a short-term memory problem may not remember the mathematical steps in order, or forget certain problem steps. These students may remember most of the lecture words but get the words mixed up with other words or have gaps in their memory. These students may not be able to remember facts, understand concepts and write down this information all at the same time. Mathematics notes on problem steps may be in the wrong order or absent.

Students with LD having auditory processing problems may have good short-term memory but can not be able to process all the words in the lecture. These students will misinterpret part of the instructor's lecture resulting in gaps in their lecture notes. These same students may also write down misunderstood words that make no sense at all. In either case, these students have difficulty understanding the lecture and recording notes.

Some of the learning accommodations for students with LD having short-term memory or auditory processing difficulties are note-takers, tape recorders, physical proximity, phonic ear, computers and video tapes. The physical proximity accommodation means that the student should sit in the center of the class and as close to the front as controllable. The phonic ear is a wireless FM receiver worn by the student, the transmitter by the instructor. The phonic ear reduces distracting sounds that interfere with learning.

These processing disorders have focused on the student receiving information to be stored in the brain. Once the information is stored in the brain, then another set of processing functions are involved in recalling the information or analyzing it. Think of the brain as a computer that has several ways to input information in its memory. Information could be learned via the key board, a scanner, modem, computer disk, or direct auditory input. If these input devices are slow or defective, then the computer, like the brain obtains defective information. The brain has difficulty remembering or analyzing this defective information resulting in ineffective learning. Then it becomes, "Garbage in, garbage out," as computer programmers say. Learning accommodations are to help the brain receive the correct information through different input devices such as note-takes, tape recorders and tutors.

Severe Processing Disorders

Long-Term Retrieval/Long Term Memory

Some students have difficulty with mathematics due to long-term retrieval/long term memory problems. Students may appear to know their basic mathematics skills one day, and then several days later, forget how to do basic calculations. These students may score good grades on quizzes but fail major tests. When they are taught a forgotten concept again, they can relearn it faster than the first time. This means that they retained some of the concept knowledge but not enough to remember the total concept. These long-term memory/retrieval problems are different from thinking/reasoning problems.

A long-term memory/retrieval problem usually applies to the mechanics of doing mathematics problems, instead of not understanding a concept. The student can understand the logic or reasoning behind the concept, and after being reminded how to do the steps, can work the problems.

If your students have a long-term memory/retrieval problem, then some learning accommodations can be mathematics video tapes, mathematics study skills, note cards, calculators, computer software, fact sheets, and tutors. These students will have to constantly review difficult mathematics procedures to keep from forgetting the correct procedures. Calculators should be used to take the place of the mechanical aspects of mathematics. Fact sheets (Reference A) can contain important mathematical properties and formulas with examples. However, even with these learning accommodations, some students with LD will not have the ability to maintain enough mathematical knowledge to become successful in their mathematics courses. Remember, learning mathematics is linear.

Fluid Reasoning

Students with LD having fluid reasoning (abstract) problems will have extreme difficulty learning mathematics. This could also include students with head injuries. These students usually have poor organizational skills, poor problem solving skills, and trouble understanding causal relationships. The difficulties causing the most problems are poor abstract reasoning and difficulty generalizing from one experience/idea to new situations. If some of your students have this type of learning disability, they may learn how to do a mathematics problem but not be able to generalize that concept to their homework problems. They may demonstrate the knowledge of a mathematics concept one day and forget how to use the same concept the next day. This usually occurs because they are trying to memorize mathematics problem patterns, because they cannot understand the mathematics concept. It may be difficult for these students to learn mathematics, but these students should be helped, so that they may go as far as they can in mathematics.

Learning accommodations for students with LD having thinking/reasoning problems are extensive. Most of these students will need a combination of note-takers, tape recorders, handouts, mathematics video tapes, fact sheets, tutors and calculators. Tutors who are trained in helping students with LD are the most important learning accommodation. Tutors must start working with these students no later than the second week of class.

These students need to use fact sheets which are similar to the ones used by students with long-term retrieval/long-term memory processing disorders. These fact sheets should also have information on how to work the problems in the students own words. There will be a need for more examples to cover each different concept to be tested. There could be five to ten pages of fact sheets developed as learning accommodations. An example of learning accommodation fact sheets are in Reference A. Reference A is the actual learning fact sheets developed by a tutor and his student.

Calculators are another learning accommodation for students with LD who have fluid reasoning problems. Calculators can reduce some of the problems with the mechanics of mathematics. This will allow the student to have a better focus on mathematics concepts. Also programmable calculators can be used to remember the formulas and calculate the answers.

Computer software programs can be used to do some of the abstract reasoning and computations for students. Students can put the numbers into the formulas and have the computer analyze and solve the equations. The students can then interpreted the results and apply it to real life situations. Students are already using these procedures for statistics courses and calculus courses.

This discussion has focused on learning accommodations for severe processing disorders. Students with LD may have problems in one or more of the processing areas. This means a combination of learning accommodations may be needed to enhance their learning. A more extensive explanation of processing disorders on learning accommodations is in **Math and the Learning Disabled Student: A Practical Guide to Accommodations** (Nolting, 1991).

Severe processing disorders focus on retaining or analyzing the information received to that area of the brain. Think of the brain as a computer that has a good central processing unit but has a few bad chips. These bad chips only have the capacity to process a limited amount of information. For this example, the "bad chips" are in long-term retrieval, long-term memory, or fluid reasoning. Through accommodations we try to reroute the these functions around the "bad chips" to other parts of the brain. The other parts of the brain can compensate to a certain extent but the processing signal keeps returning to the "bad chips". No matter how often or in what direction we try to send the signal through the "bad chip", the information only computes to a certain level and then stops. Continuing to send the single over and over again can burn out the "bad chips".

Instead of throwing away the computer, we replace the bad chips or use the computer for another purpose. Remember, the computer has a good central processing unit and other good chips. Unfortunately we can not replace the student's "bad chips", but we don't want to throw away the student either, especially if the student has a super chip in English or in other non-mathematics dependent fields. This is when we can substitute the mathematics course(s) since the student will not need to use those "chips" in their major or job field.

Course Restructuring

Due to the fact that mathematics is a linear learning process, students with severe processing disorders in most cases can not learn all the required course material. This is especially true of students who have to take a series of algebra courses to graduate. Some postsecondary institutions have customized their course curriculum to match the cognitive abilities of the student. Instruction on an individual bases teaches algebra concepts that can be understood. The instructor builds on theses concepts, until the student has maximized his/her ability to understand that concept. When students cannot expand or generalize that concept, a new concept is introduced. For example: A student may be taught how to solve an equation with one variable but can not understand how to solve an equation with two variables. The instructor would not continue to work on solving equations with two variables but would teach a new concept. The new concept might be teaching students how to graph equations. The student is then graded on a different set of standards compared to the other student.

However, even with good instruction, some students can not master all the course content. The mastery of course content also depends on the difficulty level of mathematics required for graduation. Students can reach a point in their course that requires previous algebra concept knowledge which have not been learned. Continuing to have these students repeat failed courses due to previous unlearned concepts is detrimental to the student, the instructor and the institution.

Instructor Ability

Depending on just getting the right instructor to teach the student mathematics concepts is tentative at best. The improvement of mathematics curricula has helped to enhance instruction, but the instructor's ability to teach depends upon their own individual skills. A survey indicates that 67 percent to 90 percent of instructors do not report competence or familiarity with nontraditional theoretical or instructional approaches to teaching algorithms (Carpenter, 1985). However,

Peterson, Capenter, and Phoneme (1989) indicated that an instructor's knowledge of problem-solving or number-fact strategies did not predict students' achievement scores. The best predictors for math achievement was the instructor's knowledge of individual student's problem solving skills. These problem skills would include understanding the processing deficits of the student and how to teach to the student's learning strengths.

Mathematical instruction is a major part of special education in middle school and high school, requiring as much as one-third of the instructional time for students with LD. Even this amount of instructional time has not successfully helped students with LD learn mathematics. Teachers must go beyond the traditional textbooks and materials to effectively instructing students with LD. However, the major problem is the lack of appropriate mathematics instruction procedures for teaching students with LD (Zentall & Frekis, 1993). This is especially true with postsecondary mathematics instructors who are excellent content specialist but do not have the skills or background of special educations teachers.

Tsetse and Gerber (1993) focus on the individual difference of community college students with LD. They suggest that nondisabled students require only regular instructional procedures, but students with LD must have deliberate and elaborate instruction to learn. Unfortunately, current literature on this subject is only tentative at best and especially lacking in the area of learning word problems. They evaluated the literature on mathematic and students with LD by stating, "In sum, despite promising theoretical ideas, our knowledge about attributes of older students, and what instructional interventions strategies might prove most effective, remains preliminary at best." College mathematics instructors have limited if any background in teaching students with LD and little research for assistance.

It appears the best way to help adult students learn mathematics is to understand their individual learning needs and limitations. This should replace the practice of having students repeat courses with different instructors just to see if their mathematics learning improves. However, even these suggestions will not help every student with LD complete all required mathematics courses.

Mathematics Study Skills

Most students with LD have additional problems in time management, test anxiety, study skills, and test-taking. To be academically successful, students with LD must focus on these additional learning problems as much as their learning accommodations. Montague, Bos, & Doucette (1991) in their study of students with LD indicate that affective attributes is one of the major factors that effect student learning and must be considered when designing instruction. The other major factors effecting mathematics learning are cognitive and metacognitive.

Hoover (1993) expands the idea of teaching study skills to students with LD and their parents by stating, "Study skills are essential in the overall learning process, whether students are in the early elementary grades or in high school." Study skills are not typically developed without direct instruction. To maximized students' ability to learn, study skills must be directly taught to students.

Norton (1992) conducted research on the study habits of college students with LD. The results indicated that students with LD needed more assistance in math, spelling, writing, and reading comprehension. Students with LD compared to nondisabled peers did not understand their notes as well and had more difficulty understanding and completing assignments. These students needed to be taught note-taking strategies and listening skills. Diagnosing other specific deficiencies, such as determining if math concepts or careless errors are the main problems in learning mathematics, needs to be conducted.

Vogel, Hruby, and Adelman (1993) go one step further to reduce the failure of students by challenging disabled student service providers in assisting students in understanding their LD. A students' problems must be individually diagnosed and analyzed for remedial programs and learning strategy deficits, and then taught these skills in order to become successful in postsecondary education.

One solution to mathematics study skills training is the text, **Winning At Math** (Nolting, 1991), and the computer program, **Math Study Skills Evaluation** (Nolting, 1991). The computer program analyzes the students mathematics study skills, prescribes effective learning strategies, and suggests specific page numbers to read in **Winning At Math** (Nolting, 1991). The computer program also has specific accommodation suggestions for students with LD. In addition to mathematics study skills, the text has a separate reference section for students with LD. Teaching students with LD mathematics study skills can improve their mathematics learning up to the point where their severe processing deficits interferes with any additional learning.

Testing Accommodations Based on Processing Disorders

Appropriate testing accommodations for students with LD are used to separate measuring a student's learning disability and measuring the student's mathematics knowledge. This is especially true when time is considered a major factor in measuring mathematics knowledge. In the past, when some students with LD were given a mathematics test, the test measured both the degree of their learning disability and their mathematics knowledge. In almost all these cases, the result was a lower mathematics grade which was not a true indicator of the student's mathematics knowledge. This grade frustrated students and in many situations frustrated mathematics instructor. The student was frustrated, since he/she knew more than the test indicated. The mathematics instructor also became frustrated, since the student demonstrated mathematics knowledge in class but failed the test. This section will give a brief review of appropriate testing accommodations for mathematics students with processing disorders. For a more extensive review of this subject matter consult, **Math and the Learning Disabled Students: A Practical Guide for Accommodations** (Nolting, 1991).

Timed mathematics tests will cause major problems in determining mathematics knowledge. Students with LD who have visual processing speed/visual processing disorders need as much extra time as possible to compensate for their learning disability. Depending on the type and severity of their learning disability along with the type of test, these students may need one - and - a half to two times the normal test time. However, some students have used up to four times the normal time, if they have a severe processing problem and are taking mathematics tests with graphing problems. Additional testing accommodations for students with visual process speed/visual processing problems are enlarged tests, test readers, test proof readers, audio recorded, calculator and private testing area.

Test accommodations for learning disabled students with auditory processing/short-term memory problems depend on the type and length of the mathematics test. Students may need extended time, a private testing area with a test reader or tape recorder. Students with LD may have to read the test questions many times or return to different parts of the questions to obtain a full understanding.

These are a few testing accommodations for learning disabled students with processing disorders. This does not mean that students with LD will need all these testing accommodations to circumvent their learning disability. The combination of testing accommodations usually recommend is extended test time, enlarged test and a private testing area.

Testing Accommodations Based on Severe Processing Disorders

Students with LD having fluid reasoning problems, long-term retrieval or long-term memory problems will need special testing accommodations. It will take these students longer to remember how to work the problems or to do the mechanics of working the problems. These students will need all the previously suggested testing accommodations and fact sheets.

Using calculators on tests are very important; calculators increase the chances of using the correct concept by decreasing the chance of careless errors. Fact sheets are needed to help students remember the formulas and proprieties in order to work the problems. The fact sheets can be the same that were developed as part of their learning accommodation. The fact sheets could have at least one example on each concept to be used as a guide to working the problems. Reference B has copies of some of the exact fact sheets use for an intermediate algebra final exam. These facts sheets were typed and reduced from 14" X 17" size because the student had the visual processing problem. Additional testing accommodations can be open book tests, lecture notes, homework assignments, previous tests, programmable or graphing calculators and computers.

Alternate test forms can be used for students that were taught through course restructuring or have difficulty taking traditional mathematics tests. Traditional test forms will cause difficulty for certain students with LD. Even though large print tests and audio cassette tape test forms have been discussed, there are other test forms that could be used. Some of these other test forms are oral tests, video monitor enlarged tests, computer enlarged tests, computer speech enhanced tests, and take home tests. Each test form has its advantages and disadvantages for the student and instructor.

Students who were taught through course restructuring will need individual oral or written tests to measure their knowledge. The oral test could have the student do the problems and explain the steps to the instructor. The instructor would then ask the students questions pertaining to working the problems and understanding the concepts. The written tests could have several different types of problems that the student could select based on the individual taught material. The test questions would have to be based on the material taught to the students instead of regular course material.

Learning disabled students need to be taught about the effects of their learning disability on testing. They need to be able to explain to their instructor how testing accommodations can circumvent some aspects of their learning disability and measure their true mathematics knowledge. Most mathematics instructors will try to understand the student's learning disability and arrange the testing accommodations. However, Disability Support Service providers, with the permission of the student, have the responsibility of making sure the mathematics instructor understands the students disability and is informed about the most appropriate testing accommodations.

Chapter 4

Eligibility Determinationss for Course Substitutions

4

The process for determining the eligibility for a math course substitution usually follows several guidelines. Some postsecondary institutions have stated guidelines to follow, while other institutions have not even considered a course substitution policy. Disability Support Service providers must become aware of written requirements as well as the philosophic attitudes of their institution towards math course substitutions. If your institution does not have a formal course substitution policy then find out the history of previous course substitution requests. Understanding the results of previous student course substitution requests will give you insight on how to proceed.

Qualified Disabled Individual

Disability Support Service providers need to educate the administration and mathematics faculty about the purpose and need to grant math course substitutions. They must be prepared for administration and mathematics faculty questions. This education can start with understanding the definition of a "qualified disabled individual" and what Section 504 says about academic adjustment under 34 CRF 104.44.

The reason for academic adjustments is to allow the student to maximize his/her chance to learn mathematics and to demonstrate that learning. A student's math achievement can be enhanced through appropriate accommodations, study skills training, and student effort. Achievement can be documented through instructional reports and grades. Student achievement in other courses can also be documented through grades and over all GPA. This is the first step in helping students with LD obtain course substitutions.

The definition of a student with a learning disability was explained in Chapter One. The definition of a "qualified disabled student" may differ, depending on the college and program of study. Students must be designated as a "qualified disabled student" just to apply for the course substitution. According to Section 504 34 CFR 104.3 (k) (3) "qualified disabled person"is:

> *With respect to postsecondary and vocational education services, a disabled person who meets the academic and technical standards requisite to admissions or participation in the recipient's education program or activity.*

According to the Americans With Disability Act a qualified individuals with a disability is:

> *... any individual with a disability who, with or without reasonable modifications to rules, policies, or practices, the removal of architectural, communication, or transportation barriers, or the provision of auxiliary aids and services, meets the essential eligibility requirements for the receipt of services or the participation in programs or activities provided by a public entity.*

If your institutions indicates that the student is not a "qualified disabled student", then the student cannot apply for a course substitution. Some postsecondary institutions have claimed that students were not learning disabled or were not a "qualified disabled student", thus denying access to course substitution procedures. One college claimed that a student who was applying for the course substitution was not learning disabled, since the college labeled her as having a learning problem. Even though she was assessed by the college learning specialist as being LD and had received accommodations, the college insisted that she only had a learning problem and was not disabled. By labeling her as having a learning problem, she was denied access to the course substitutions procedures.

When the Office of Civil Rights grievance was filed, the first procedure of the investigation was to establish if the student was a "qualified disabled person". This is a two-step process. Using the Section 504 guidelines discussed in Chapter One, the student was receiving accommodations and being treated as disabled. By the fact she received accommodations and was treated as disabled made her a disabled person.

The second part was to rule if the student was a "qualified disabled person". The student was ruled as being a "qualified disabled person" since she was admitted to the college in question and met the standards necessary for admission while participating in the college's program for over 10 years. Under these rulings, the student was deemed eligible to apply for a course substitution. Disability Support Service providers must be aware of their institution's definition of disability or "qualified disabled person" to help students in the substitution process.

Academic Adjustments

The Section 504 regulations do mention course substitutions but do not mention waivers of courses. Section 504 CRF 104.44 states:

> *A recipient shall make such modifications to its academic requirements as are necessary to ensure that such requirements do not discriminate or have the effect of discriminating, on the basis of handicap, against a qualified handicapped applicant or student. Academic requirements that the recipient can demonstrate are essential to the program of instruction being pursued by such student or to any directly related licensing requirement will not be regarded as discriminatory within the meaning of this section. Modifications may include changes in the length of time permitted for the completion of degree requirements, substitution of specific course required for the completion of degree requirements, and adaption of the manner in which specific course are conducted.*

Most service providers try to avoid any discussion of waivers, except in very rare situations. The discussion of wavers can weaken your case for not taking the mathematics course(s) by appearing to "water down the curriculum" and undermining the emphasis of equal treatment under the regulations (Jarrow, 1992). The best way to approach your institution about students with difficulty learning mathematics is to suggest course substitutions instead of course waivers.

Jarrow (1992) indicates that Section 504 does allow course substitution as an appropriate modification to be considered as an academic adjustment. Jarrow continues:

> *Substitutions would not be required if the course or content is found to be essential to the area of study for the student and if making a substitution would require a substantial change in an essential element of the curriculum. It is the responsibility of the institution to show that a requirement is essential to a given course of study and that making a substitution would substantially alter the curriculum for the student. Just indicating that a given course is required of all students is not enough to establish it as essential to the curriculum. Section 504 does not require that the institution make substitutions in course requirements on request, only that it be willing to consider substitution as a possible form of academic adjustment on a case-by-case basis.*

Institutions in the past have denied academic adjustments to students with LD by performing cursory reviews or by just saying "no'. However, the court of appeals ruling in the Wynne vs. Tufts University School of Medicine (1992) indicated that the university failed to adequately support it's position that altering its test format would fundamentally alter the program of study. This ruling is very important because it places the burden of proof of denying academic adjustment (accommodations, course substitutions) with the institution (Heyward, Lawton, and Associates, 1992b). Another important case is the Office of Civil Rights (OCR) ruling in the Southwest Texas State University vs. OCR (case no. 06902084). In this case a student with LD was denied a mathematics course substitution request. The university was found to be in violation of Section 504 because it did not provide evidence that the substitution was a substantial alteration of the student's bachelor's degree (Kincaid, 1992). Institutions must now provide adequate documentation when denying course substitution.

Your institution must be willing to hear course substitutions requests from its disabled students. It is part of the service provider's responsibility to provide the evidence for the course substitution and make a recommendation. If the course substitution is denied, written documentation by the institution must prove that the mathematics course was an essential part of the student's educational program.

When a student applies for a course substitution, the Disability Support Service Provider needs to be prepared to answer questions about the student's background. Using the information in Chapter Two, the services provider can explain the reasons for the student's mathematics learning disability. Make sure that you quote the correct resources, explain the actual processing deficits, and it's effects on learning mathematics. Next you can use the information in Chapter Three to explain the appropriate learning and testing accommodations given to the student. The accommodations need to be so inclusive that additional accommodations would not be appropriate or would not improve the student mathematics learning or testing results. Being unable to explain the nature of the student's processing disorder or not being able to assure that the student has received all appropriate accommodations can be reasons for denying a course substitution.

The student's mathematics study skills training and background can be a subject of questions. The service provider needs to show that the student has been using appropriate math study skills. Then the mathematics instructors can not suggest that student's mathematics course failures were due to ineffective or inappropriate study skills.

Student effort is a major area of concern when math course substitutions are requested. Service providers must be able to show that the student has tried (ie., made a good faith effort to succeed) in the course and has used available resources. The student might want to keep a study log and home work assignments to validate his/her learning effort. I had several students keep logs showing that they studied 30 to 40 forty hours a week on mathematics and still could not pass. The logs were very impressive. Logs on tutor time and time spent with the instructor should be kept. Letters from the math lab tutors and instructor, indicating that effort was put forth, are very valuable. Depending on the institution, the student might need all or only some of this data to support sufficient effort.

Questions will always be raised about the ability of the student to pass his/her other classes. You need to know the student's GPA, and whither the student has any special talents or academic skills. The numbers of "A" and "B" in other classes can be listed to show that the student has the ability to learn in other subject areas. The total number of hours attempted and passed is also helpful. Letters from instructors, indicating the student has the skill to be successful in his/her major area without mathematics, is very helpful. Knowing how many courses the student has to graduate, especially if math is the only course left, is excellent evidence of effort and ability. This also shows that the mathematics course is not essential as part of the educational requirements.

If a student has low grades, then expect questions regarding the reason for the lower grades. If the student has borderline ability, then you need to feel comfortable in your own mind that the student has the ability to be successful in college. This is especially true of "open door" institutions that have students with all ranges of intelligence. If the student has low grades and borderline ability, then it becomes difficult to recommend a math course substitution. Remember, you, too, are being evaluated when recommending course substitutions.

Student Questions

After understanding the questions that the administration and faculty might have we can focus on the student's questions. Usually the student's first question is, "How many times do I have to fail the math course before having a good chance of getting a math substitution?" The second question is "Does the institution have to grant math substitutions?" The answer to the first question depends on the institution's policy and philosophical orientation. Students should not have to fail a mathematics course before applying for a course substitution. If the student has documentation verifying serious deficits in information processing abilities related to the course then student should be allowed to request a course substitution. The student should also had appropriate mathematics accommodations in high school or in previous mathematics courses. Adhearding to a fail first philosophy impacts the student's academics status and damages self-esteem and motivation (Brinckerhoff, et. al., 1993). However, institutions usually have the student try the mathematics course at least one time befor requesting a subsitution.

The answer to the second question is "No". Remember, course substitutions are made on an individual basis depending on whether such substitutions would require a "substantial change in an essential element of the curriculum". In other words, would the substitution causes problems in taking other courses in the student's major or in some cases obtaining employment? This "essential part of the program" question is the major determination in a ruling on a course substitution should be granted.

Section 504 does not state anything about the number of times you have to take a mathematics course and fail it before requesting a substitution. Your college might have its own formal or informal policy on how many times a student has to fail math before requesting a substitution. I have known students who have failed different mathematics courses several times before requesting a substitution, while other students have requested a math substitution without taking any college math courses.

The real question is, "Did the student fail mathematics while obtaining all the appropriate learning and testing accommodations?" If the answer is, "Yes", then it is my opinion that the student has learned as much math as possible and does not have to continue to fail math to prove his/her inability to learn mathematics. This is why it is important to give the student the most appropriate accommodations as soon a possible. The student does not have to receive all the accommodations suggested in Chapter Three, only the ones deemed appropriate by the institution.

In some cases, students have been granted math course substitutions without taking any mathematics classes. These cases of substitutions are rare but do exist. In one case, a major university in the southeast granted a incoming freshman a course substitution without taking any mathematics courses. The decision was made due to the overwhelming documentation that the student had learned as much algebra as she could in high school and would not pass the lowest algebra course. This student, while in high school, had already received math study skills training and appropriate learning and testing accommodation. Her main processing disability was abstract reasoning. The documentation validated that trying to teach her algebra courses above her ability would be futile and could be psychological damaging. Since the mathematics courses were not an essential part of her major, the university substituted other subjects for math. Based on similar documentation, community colleges have also substituted mathematics courses before the student attempted a mathematics course.

Showing that the mathematics course is not an essential part of the program is crucial in obtain approval for a mathematics course substitution. A student's major is a prime consideration in determining if a course substitution will be granted. Jarrow, in her text, **Subpart E: The Impact of Section 504 on Postsecondary Education** (Jarrow, 1993), gives an example of two students who requested mathematics course substitutions. One student was majoring in Recreational Therapy and the other student was majoring in Business. Both students had above 2.0 GPA's and requested a substitution for college algebra. The student majoring in Recreational Therapy was granted the course substitution, and the Business major was not granted the course substitution. The committee felt that for the Recreational Therapy major, college algebra was not an essential part of the curriculum. However, college algebra was an essential part of the curriculum for the Business major. College algebra was a prerequisite to other required courses in the business major. This is a good example of case-by-case decision making required by Section 504.

Disability Support Service providers need to establish that the mathematics course requested for substitution is not an essential part of the student's curriculum. If the student does not require math skills in any remaining courses, then math course is not needed for their program. If the student has to take courses with math prerequisites, then information must be presented to indicate that the student can pass the course without the prerequisite math skills. To add credibility to the request, obtain letters from instructors in the student's major supporting the course substitution. The letters should indicate that the student has sufficient math skills to complete the major and become employed. Letters from employers, indicating that students with similar math skills are employed in the student's major are supportive. Students who are in community colleges should try to obtain a letter from the university they are transferring to, indicating that they would substitute the math courses or that the student already has sufficient math skills for their major. Showing that the student does not need anymore mathematics skills to be successful in their major and can be employed after graduation is the main key to course substitutions.

Chapter 5

Course Substitution Procedural Suggestions

5

Some colleges have developed formal and informal course substitution procedures. In some states like Florida, the state legislators have required all state colleges and universities to develop course substitutions procedures. Each postsecondary institution needs to develop formal course substitution procedures in order to be consistent in their decisions. If the Office of Civil Right has a grievance filed against your college regarding course substitutions, they will first look for violated college procedures and inconsistencies in granting course substitutions. To protect your institution, course substitution policies/procedures need to be approved by the Board and administered consistently. It will be difficult to prove that the institution did not discriminate against a student by denying a course substitution without having such a policy in place.

The fist step in developing a course substitution policy/procedure is to locate any state rules that refer to such procedures. Florida has such a rule and states:

6H-1.041 Substitute Admission and Graduation Requirements.

(1) Each district board of trustees of a public community college shall develop and implement policies and procedures for providing reasonable substitution for eligible students as required by Chapter 86-194, Laws of Florida. In determining whether to grant a substitution, documentation to substantiate that the disability can be reasonably expected to prevent the individual from meeting requirements for admission to the institution, admission to a program of study, entry to upper division, or graduation shall be provided ...

(2) The policies and procedures shall include at least the following:

 (a) A mechanism to identify persons eligible for reasonable substitutions due to vision impairment, hearing impairment, dyslexia, or other specific learning disability.

(b) A mechanism for identifying reasonable substitutions for criteria for admission to the institution, admission to a program of study, entry to upper division, or graduation related to each disability.
(c) A mechanism for making the designated substitutions known to affected persons.
(d) A mechanism for making substitution decisions on an individual basis, and
(e) A mechanism for a student to appeal a denial of a substitution or to appeal a determination of ineligibility.

(3) The policies shall provide for articulation with other state institutions which shall include, at minimum, acceptance of all substitutions previously granted by a state postsecondary institution.

(4) Coordination of the provision of technical assistance in the implementation of this rule shall be provided by the Division of Community Colleges in conjunction with the State Department of Education.

(5) Each community college shall maintain records on the number of students granted substitutions by type of disability, the substitutions provided, the substitutions identified as available for each documented disability and the number of requests for substitutions which were denied.

The Florida rule requires community colleges to develop policies and procedures to allow students with learning disabilities to request course substitutions. The four year universities have a similar rule requiring the development of policies and procedures. The community colleges are allowed to develop their own policies and procedures, as long as the state rule is met. The State Board of Community Colleges has further interpreted the provisions, and they are located in Reference C.

Samples of College Substitution Procedures

Some of the community colleges developed abbreviated substitution graduation requirements. The draft proposals are divided into seven sections from "Mechanism For Identification of Persons Eligible" to "Additional Comments". For guidance on developing your own procedures six sample draft proposals are below:

College A

Mechanism For Identification of Persons Eligible

Self identification.

Mechanism For Identifying Reasonable Substitution Criteria For Graduation

Three attempts with receipt of failing grade, if it does not constitute a fundamental alteration in the nature of program. Dean of Students consults with Dean of Academic Affairs to determine an adequate substitute course based on campus availability. Disabled Coordinator assists.

Mechanism For Making Designated Substitutions Known To Affected Persons

Outlined in the college procedures manual.

Mechanism For Making Substitution; Decision On An Individual Basis

Three attempts with receipt of failing grade and a 2.5 GPA in other required courses, counsel individual, student interview.

Mechanism For Students To Appeal Denial Determination

Regular college appeal process as outlined in college procedure manual.

Articulation Provisions

Information about policy will be disseminated to local upper level colleges and universities.

Additional Comments

None

College B

Mechanism For Identification of Persons Eligible

Self identification.

Mechanism For Identifying Reasonable Substitution Criteria For Graduation

General guidelines established by the substitution committee.

Mechanism For Making Designated Substitutions Known To Affected Persons

Not indicated.

Mechanism For Making Substitution Decision On An Individual Basis

A substitution that has not been previously approved, with the instructor's permission may be submitted to the substitutions committee which meets at the request of the Disabled Student Services Coordinator.

Mechanism For Students To Appeal Denial Determination

The Student may pursue adjudication by (via) the Committee on Academic Review and Petition. Complaint must include the resolution sought.

Articulation Provisions

The substitution committee will approve the incorporation of policies from transfer institutions. This includes the acceptance of all substitutions previously granted by a state postsecondary institution.

Additional Comments

None.

College C

Mechanism For Identification of Persons Eligible

Responsibility of the office of Student Services. Eligibility will not be assumed or assigned with documentation of the disabling condition. Documentation will be obtained with the student's consent and maintained confidentially by the Office of Disabled Student Services.

Mechanism For Identifying Reasonable Substitution Criteria For Graduation

Shall be determined by a Committee which will include at least the eligible student's advisor, the Coordinator of Disabled Student Services, and the appropriate division chairman.

Mechanism For Making Designated Substitutions Known To Affected Persons

Students will be notified of acceptable substitutions by their division chairman. In orienting affected students to the services available to disabled students, they will be informed of the procedures that exist for making reasonable substitutions.

Mechanism For Making Substitution Decision On An Individual Basis

Students who need and are eligible for substitutions will request such through their faculty advisor. Identification of reasonable substitutions shall be determined by a committee which will include at least the eligible student's advisor, the Coordinator of Student Services; and the appropriate division chairperson.

Mechanism For Students To Appeal Denial Determination

Students will be informed of their rights to appeal the denial of a substitution or the determination of ineligibility during their orientation by a Counselor of Disabled Student Services (office). Appeals will be handled through existing student grievance procedures.

Articulation Provisions

The college will actively work with other postsecondary institutions to assure uniformity of substitutions that are provided and acceptance of substitutions previously granted by their postsecondary institutions.

Additional Comments

Faculty and staff will also be made aware (via workshops and handouts) of policies and procedures in order to assure that all affected students are adequately advised about substitutions

College D

Mechanism For Identification of Persons Eligible

A written statement summarizing the substitution policy is to be available in all departments which provide direct services to students. Also, voluntary self-identification of students with specific disabilities or handicaps is to be included with registration information. Referrals are to be made to handicapped services for eligible persons who require assistance meeting courses, programs, and/or graduation requirements.

Mechanism For Identifying Reasonable Substitution Criteria For Graduation

The relationship of each admission/graduation criterion to state-mandated requirements will be justified by the program chairperson or department chairperson. Each admissions criterion and/or reasonable substitution must be approved by the program chairperson, the appropriate dean and the director of admissions. Appropriate substitutions for eligible persons for graduation and general education requirements are to be recommended by department chairpersons where there is reasonable opportunity for successful employment or meeting admissions requirements for entry into upper division program.

Mechanism For Making Designated Substitutions Known To Affected Persons

A written statement summarizing the substitution policy is to be available in all departments which provide direct services to students and will be include in registration information.

Mechanism For Making Substitution Decision On An Individual Basis

The eligible person will be referred to the department chairpersons where there is reasonable opportunity for successful employment or meeting admissions requirements for entry into upper division programs.

Mechanism For Students To Appeal Denial Determination

Appeals of denial of a substitution or determination of ineligibility are made to the academic appeals committee.

Articulation Provisions

Not addressed.

Additional Comments

None.

College E

Mechanism For Identification of Persons Eligible

Self identification and written documentation of disability. Voluntary Student Disability Survey which is used by the college coordinator of services for students with disabilities to contact students who may be eligible for substitutions.

Mechanism For Identifying Reasonable Substitution Criteria For Graduation

The Dean of Academic Affairs will be notified of all request for substitutions, and with the cooperation of the appropriate chairperson, Coordinator of Services for Handicapped Students, Director of Admissions and Records, Dean of Student Development and appropriate instructional personnel will seek to develop a program of study containing reasonable and appropriate substitution.

Mechanism For Making Designated Substitutions Known To Affected Persons

Prospective students will be advised that the college encourages voluntary self identification of any disabilities or handicaps that may prevent them from meeting requirements for admissions or graduation and that support services are available to meet these special needs. Students requesting substitutions will be contacted personally and advised on the actions taken by the committee.

Mechanism For Making Substitution Decision On An Individual Basis

Committee review on an individual basis subsequent to request in writing and acceptable documentation.

Mechanism For Students To Appeal Denial Determination

Via established student grievance procedure as outlined in "The Students Rights and Responsibilities" section of the student hand book.

Articulation Provisions

The college will accept substitutions previously granted by other Florida postsecondary institutions. Documentation should be requested by the transferring student and sent from the granting institution.

Additional Comments

None.

College F

Mechanism For Identification of Persons Eligible

Disabled Student Identification Form to be distributed in orientation folders and student completion of substitution request form.

Mechanism For Identifying Reasonable Substitution Criteria For Graduation

Proposal that request be submitted by the Handicap Services Coordinator, along with a recommendation to the Academic Affairs representative and the Registrar's office for review.

Mechanism For Making Designated Substitutions Known To Affected Persons

Proposed listing of designated substitution allowances and procedures in the college catalog under graduation requirements for the general knowledge of all affected students.

Mechanism For Making Substitution Decision On An Individual Basis

Proposed evaluation of the needs of each individual student on an individual basis, allowing him/her the greatest potential for success.

Mechanism For Students To Appeal Denial Determination

Substitution denials may be appealed through the student services and Academic Affairs Department Heads for further review and recommendations.

Articulation Provisions

Proposed one year review period to gather information, research legalities and coordinate substitutions criteria between the College and Public School System. Also, design of an articulation agreement with the State University System to insure a smooth transition into the upper divisions at state universities.

Additional Comments
None.

These sample abbreviated draft substitution proposals were written for discussion purposes only and have probably been changed several times. After reviewing these draft proposals you should have a better idea of the procedures which can be included in your substitution policy or guidelines.

Actual College Substitutions Procedures

Figure 1 is an example of a substitution rule developed in accordance with Florida State Board of Education Rule 6A-10.041. The procedure requires the substitution request be made to the President of the College and is a two step process. First, a panel determines if the request is valid and sends the decision to the President within 15 days of the student's request. This panel is made up of student services representatives, academic representatives and the 504 coordinator. If the President approves the request for substitution of requirements the next step is for another panel to determine whether a reasonable substitution exists that does not constitute a significant alteration in the program. The panel makes its recommendation within 15 days to the President. If the student is denied the course substitution, then an appeal can be made to the Board of Trustees. This substitution rule has a built in time line so that the substitution process doesn't drag out over a long period of time.

Figure 2 is another substitution policy which requires documentation that the disability can be expected to prevent the individual from meeting the graduation requirement. This means that the student must provide documentation that the processing disorder is causing difficulty in learning mathematics. The substitution request is then forwarded to the Vice President of Academic Affairs, instead of going to the President. The substitution committee is appointed by the Vice President of Academic Affairs and is not designated. The committee decides if this is a valid request and can make a recommendation of a course substitution to the Vice President. The Vice President makes the final decision on the course substitution. If the student is denied, then the appeal goes to the President.

Figure 2 does not require the Disability Support Service Coordinator to be on the committee, the committee decision is not binding, and the appeal does not go to the College Board. In my opinion, the Disability Support Service Coordinator (504 coordinator in some cases) must be

College Substitution Rule

In accordance with SBE Rule 6A-10.41, the following guidelines will be in effect:

A. **Eligibility Requirement:** Students requesting a substitution of requirements for graduation or admission to a program based upon hearing impairments, visual impairments, or specific learning disabilities, must submit documentation from competent professionals verifying that these disabilities exist and to the degree specified by Florida Statutes.

B. **Request for Substitution of Requirements:** Students who request a substitution of requirements shall make the request in writing to the President. That request must identify the specific disability(ies), and course(s), and provide the appropriate documentation as stated in Part A above.

C. **Review of Requests for Substitution of Requirements:** A Panel consisting of the following people shall review the request to determine its validity, and shall make its recommendation to the President for his action within fifteen (15) working days from receipt of the Request.

- Dean of Student Services, Chairman
- Dean of Academic Affairs
- Dean of Admissions and Records
- Dean of Instruction
- 504 Coordinator

D. **Designation of Reasonable Substitutions:** If the President approves the Request for Substitution of Require ments, the following Panel, within fifteen (15) working days from receipt of the President's recommendation, shall determine whether a reasonable substitution exists that does not constitute a significant alteration in the program. The Panel shall recommend a reasonable substitution to the President or recommend that no reasonable substitution is available.

- Dean of Academic Affairs, Chairman
- Dean of Instruction
- Division Chair with Course Jurisdiction
- Two (2) Senior Faculty Members in the Department with Course Jurisdiction

E. **Approval of Reasonable Substitutions.** Final action regarding approval or disapproval shall rest with the President of the College.

F. **Appeal Process:** Students whose requests for reasonable substitutions have been denied may appeal directly to the President in writing. The President shall forward the appeal to the Chairman of the Board of Trustees who may name a panel of five (5) members of the faculty and administrative staff to review the appeal. That panel will submit its findings and recommendations to the Board of Trustees. Final action on the appeal shall rest with the Board.

G. **Records and Articulation:** The Dean of Admissions and Records shall maintain in the student record file the official approval of reasonable substitutions. These substitutions will be so designated on the student's transcript of courses. The transcript will serve as the articulation document (in the matter of reasonable substitutions) to other post-secondary institutions.

Figure 1

designated as a committee member for course substitutions requests. The Coordinator also must be allowed to bring the LD Specialist or Learning Specialist to the meeting to describe the effects of the student's learning disability on mathematics. Educating the committee members on the effects of a student's learning disability is best done by Disability Support Service services providers, since they have the best background to determine if a course substitution is warranted. It may also be legally necessary to allow such input in order to assure compliance with both the letter and the intent of Section 504.

Reasonable Substitutions of Academic Requirements

These procedures are to be followed:

1. In accordance with FS Sections 240.152 and 240.153 and the applicable rules promulgated thereunder, any applicant for admission or student who has a hearing or visual impairment or who has a specific learning disability shall be eligible for reasonable substitution of any requirement for admission, for graduation, for admission into a program of study, provided that the person's inability to meet the requirement is related to the disability and where the failure to meet the graduation requirements or program admission requirement does not constitute a fundamental alteration in the nature of the program.
   In determining whether to grant substitution of arequirement, documentation to substantiate that the disability can be reasonably expected to prevent the individual from meeting the applicable requirement(s) must be provided.

2. Persons who believe they qualify for substitutions of one or more requirements must petition, in writing, and provide documentation for such substitution. The petition for substitution of requirements must include:

   a. Identification of the specific requirements for which substitution is requested
   b. Identification of the disability which is the basis for the request
   c. Documentation that the failure to meet the requirement(s) for which substitution is requested s related to the disability.

3. A petition for substitution of one or more requirements for admission to the institution and accompanying documentation should be provided to the Dean of Admissions and Records not later than the last day to apply for admission to the College. A petition for substitution of requirement by a matriculating student should be provided to the chairperson of the division offering the program the student seeks to enter or graduate from or the Dean of Academic Affairs in the case of a person seeking the Associate of Arts Degree.

4. Such a petition for substitution of requirements shall be forwarded immediately to the Vice President of the College. The Vice President shall convene a panel to review the request and make its recommendation to the Vice President for his action within fifteen (15) working days from receipt of he petition. The panel may solicit advice and counsel fromappropriate members of the College community. The recommendation of the panel shall be in writing and shall include the rationale for the decision and, in cases where substitution is recommended, specification of the substitution. The Vice President will review the recommendation of the panel and communicate his decision, in writing, to the person submitting the petition within five (5) working days of receipt of the panel's recommendations.

5. Persons denied substitution of one or more requirements may appeal the decision to the President of the College within ten (10) calendar days of receipt of the written decision from the Vice President of the College. The President will review the original petition and documentation, the recommendation of the panel, the decision of the Vice President, and other information deemed pertinent to the appeal. The President will communicate his decision, in writing, within fifteen (15) calendar days or receipt of the written appeal. Decisions of the President with regard to the substitutions of requirements shall be final.

6. Persons who are applying for admission to the College who have documented visual or hearing impairments or specific learning disabilities and who have previously been admitted to another Florida state university or community college on the basis of substitution of requirements for one or more admission criteria and who are otherwise eligible or persons who, while matriculating at another Florida state university or community college, have successfully petitioned for and have been granted substitution or requirements at such institution; or, persons who have graduated from a Florida public high school on the basis of substitution of requirements for one or more high school graduation criteria which are also satisfied the corresponding criteria for admission to or substitution of requirements of the College.

7. The Dean of Admissions and Records shall maintain records on the number of students granted substitutions by type of disability, the substitutions provided, the substitutions identified as available for each documented disability and the number of requests for substitutions which were denied. Official approval of reasonable substitutions will be maintained in each individual student's record file. These substitutions will be so designated on the student's transcript, and the transcript will serve as the articulation document to other post-secondary institutions.

Figure 2

Substitution Admissions and Graduation Requirements for Handicapped Students

The 1973 Rehabilitation Act states that substitutions to curriculum requirements are an appropriate accommodation. In 1987 the Florida Legislature passed the 1987 Substitution Law that identified three categories of disabilities which can be eligible for substitutions. These can be substitutions of admission criteria to the institution, degree program admissions criteria and graduation requirements.

The Legislature, however, did not specify how this would be accomplished and left it up to each institution to develop and implement its' own procedures. At our institution the President designated the Provosts of each campus as his designees to grant or deny substitution requests. The Provost have formed committees comprised of counselors and teaching faculty and department personnel to consider each request on its own merits.

The procedure for a substitution request is as follows:

1. Student petitions the Office of Disabled Student Services for the substitution. The counselors work with the student to compile the necessary data that will be needed to present to the committee.. his data includes:
    a. All tests scores and interpretation of those scores showing the relationship of those test scores to the academic discipline or course.
    b. Letters of support from the teaching faculty and tutors and counselors.
    c. All transcripts showing the number of times the student has attempted the course.
    d. Overall recommendation from the L.D. Specialist.
2. This information is then presented to the committee. If the committee makes a recommendation to the provost for the substitution, they then offer several courses to be taken as substitutions. All approvals or denials for substitutions are then sent to the Provost, who makes the final determination.
3. The courses offered for substitutions for graduation requirements must come from the listing of General Elective courses for the degree. The Provost will make the determination in accordance with the students stated degree objectives.
4. All approvals are sent from the Provost office to the various departments such as the Registrar and the Disabled Student Services office and the student.
5. Denials are sent back to the Office of Disabled Student Services and to the student. Denials may result in an appeal action. This action will be through the Office of Disabled Student Services to the Provost for reconsideration. If the request is still denied then the student can file an appeal to the Student Grievance Committee. If the request is denied again then to student has the option to file a civil suit.
6. If the approval also requires modifications to curriculum or testing, then those accommodations are coordinated through the Office of Disabled Student Services.
7. This substitution becomes part of the students permanent record and is transferred to other sate institutions upon request of the student.

Figure 3

Figure 3 is a good example of a mathematics substitution procedure because it directly involves the Office of Disability Support Services throughout the process. The student's substitution request begins in the Office of Disability Support Services and the counselors work with the student to compile the necessary data. The data includes test scores, letters of support, transcripts, and a recommendation from the Learning Disability Specialist. The information is presented to a committee that consists of counselors, teaching faculty, department personnel, and the Disability Support Service Coordinator. If the course substitution is approved, then the committee offers several courses as substitutions.

Substitute Admission and Graduation Requirements for Eligible Students

In accordance with State Board of Education Rule 6H-1.041 the following policies and procedures are established by the District Board of Trustee to provide reasonable substitution for eligible students to meet admission (general admission and admission to a program of study) and graduation requirements (Chapter 86-194, Laws of Florida).

Students seeking a substitution shall present documentation to the Director of Counseling to substantiate the disability which can be reasonably expected to prevent the individual from meeting requirements. The following definitions shall apply:

a) Hearing Impairment - A hearing loss of thirty (30) decibels or greater, pure tone average of 500, 1000m, 2000Hz, ANSI, unaided, in the better ear. Examples include, but are not limited to, conductive hearing impairment or deafness, high or low tone hearing loss or deafness, and acoustic trauma hearing loss or deafness.
b) Visual Impairment - Disorder in the structure and function of the eye as manifested by at least one of the following: visual acuity of 20/70 or less in the better eye after the best possible correction, peripheral field so constricted that is affects one's ability to function in an educational setting, or a progressive loss of vision which may affect one's ability to function in an educational setting. Examples include, but are not limited to, cataracts, glaucoma, nystagmus, retinal detachment, retinitis pigmentosa, and strabismus.
c) Specific Learning Disability - A disorder in one or more of the basic psychological or neurological processes involved in understanding or in using spoken or written language. Disorders may be manifested in listening, thinking, reading, writing , spelling, or performing arithmetic calculations. Examples include dyslexia, dysgraphia, dysphasia, dyscalculia, and other specific neurological process. Such disorders *do not* include learning problems which are due primarily to visual, hearing, or motor handicaps, to mental retardation, to deprivation.

Through the orientation program, academic advisement, and registration process, students will be informed of the services for student with disabilities. On the registration form students may self-identify a disability. A counseling services brochure and a brochure describing facilities, equipment and services for disabled students will be disseminated to the district high schools and various agencies in the surrounding communities.

Substitutions which are relevant to the student's career aspirations or college major will be considered. The following reasonable substitutions may be provided to eligible students who have documented the disability in the Office of the Director of Counseling:

MATHEMATICS: OST 1324 Business Mathematics and Machines
CGS 1060 Introduction to Computer Literacy
MAT 1033 Intermediate Algebra
Science Courses

COMMUNICATIONS: SPC 1300 Fundamentals of Interpersonal Communications
JOU 2440L Literary Magazine
MAN 2130 Business Writing
Use of word processing equipment, personal computers, dictation, dictionaries and tutors

The Counseling Office will have the responsibility of making the above substitutions known to affected persons, and the Academic and Admissions Committee will study the individual cases of students who may have difficulty with the recommended substitutions listed above. The Academic and Admissions Committee will make recommendations to the Vice President of Student Services on such cases. Students may appeal to the Vice President of Student Services when the Academic and Admissions Committee denies a substitution or rules that the students is ineligible to be granted a substitution.

The College will honor substitutions granted by other post-secondary institutions in the State of Florida.

Records on the number of students granted substitutions will be kept by the Registrar and will include: type of disability, the substitutions provided, the substitutions identified as available for each documented disability, and the number of requests for substitutions which were denied.

Figure 4

The Provost makes the final decision on all approvals or denials for course substitution. If the substitution approval also requires modification to curriculum or testing, then those accommodations are provided by the Office of Disability Support Services. Appeals for a substitution denial go through the same procedure. The final appeal is through Student Grievance Committee.

The procedures shown in Figure 3 focuses on the Disability Support Services. As mentioned earlier, disability service providers must be directly involved in helping the student to present his/her case. The Office of Disability Support Services also "screens" substitution requests that do not have valid documentation which saves the college time and money. The Learning Disability Specialist's recommendation is very important and includes the reasons for mathematics learning difficulties. The first appeal goes through the same process starting with the Office of Disability Support Services. The Office of Disability Support Services can now modify learning accommodations, testing accommodations, and math study skills training. These modifications may improve the students chances of learning mathematics or present a stronger case for the next course substitution request.

Figure 4 provides another good example of a mathematics substitution policy. The student presents documentation to the Director of Counseling to substantiate the disability and that the disability is preventing the student from passing math courses required for graduation. The Director of Counseling consults with the Disability Support Service Coordinator and the mathematics instructor to discuss the course substitution. If the student has demonstrated effort and has appropriate learning and testing accommodations, the substitution is granted. The student meets with the Director of Counseling and selects a course substitution from the following courses: Business Mathematics and Machines, Introduction to Computer Literacy, Intermediate Algebra or Sciences Courses. Intermediate Algebra is a first credit algebra course and can be substituted for failing high level algebra courses that are required for graduation. These course substitutions are selected based on relevance to the student's career aspirations or college major. If the student has difficulty selecting one of the above courses, the Academic and Admissions Committee will study the individual's case and make recommendations to the Vice President of Student Services.

The difference between the policy shown in Figure 4 and other substitutions procedures is that it is mainly handled through the Student Service part of the campus. The Director of Counseling and the

Disability Support Service Coordinator look at the substitution request with input from the mathematics faculty member. In most cases, the Director of Counseling and the Disabled Student Coordinator, with input from their staff, will have the best background to make decisions about the student's learning disability. It so happens in this case that the mathematics faculty member consulted at this college is an expert in the LD field. This process is effective and valid because the experts make the decisions about course substitutions and a list of course substitutions have already been determined.

Many universities now have similar written procedures for granting mathematics course substitutions. These universities may be considering substitutions for students who transferred from community colleges without finishing their mathematics courses or Associate of Arts degree, or for their own "native" students. The transfer students may not have completed the mathematics courses because their college did not offer appropriate accommodations or their college has a history of not granting mathematics course substitutions. For such students, you need to explore whether your university offers the right level of mathematics courses. Many times these students have failed developmental or mathematics courses that are below the courses offered at your university. These students need to be sent back to the local community or junior college. Students who did not have proper accommodations but are at a mathematics level offered at the university might want to try the course again. The "native" students may have been admitted under special conditions or had high test scores in all the areas but mathematics.

Some universities have "a committee of two" who decide on mathematics course substitutions. The committee consist of the Disability Support Services Coordinator and the Vice President of Academic Affairs or another administrator who is responsible for course substitutions. Students with learning disabilities discuss their mathematics learning difficulties with the Disabled Support Service Coordinator. Before the Disability Support Service Coordinator writes a substitution request to the Vice President of Academic Affairs, he/she verifies the student's disability, measures the student's effort and reviews the accommodations. Students who are in need of developmental mathematic are asked to dual enroll at the community college. These students take their math courses with appropriate accommodation at community colleges, while being full time at the university. The student's progress is followed up with the mathematics instructor. The Disability Support Service Coordinator only requests course substitutions for students that have a legitimate request, have learned as much mathematics as possible, and in those cases where mathematics courses are not essential to their major program.

When the student meets these substitutions requirements, the Disability Support Service Coordinator writes a letter to the Academic Dean requesting a course substitution. The letter contains reasons for the course substitution and a recommendation from the Disability Support Coordinator. The course substitution may be a non-credit mathematics course that the student passed at the community college. Native students may select a philosophy of logic course, a non-programming computer science course, or courses that relate to their major. Other course substitution might have to be selected, depending on required general education courses that have mathematics course prerequisites. The Academic Dean almost always approves the course substitution and sends a memo to the records department, the student, and Disability Support Service Coordinator indicating the course to be substituted. In the future, any graduation check will list the course substitution and indicate that the student has fulfilled the mathematics requirements. An example of the correspondence letters between the Disability Support Service Coordinator and the Academic Dean is in Reference D.

This substitution procedure was reported by an AHEAD member at a major University in the southeast. She indicated that it is very successful because the best informed individuals are making the decision. This eliminates many of the problems associated with large committee decisions. She also said that most of the mathematics substitutions are made because the students have abstract reasoning disabilities. This is a excellent procedure because the Academic Dean gives the Disability Support Service Coordinator autonomy in making the decisions and realizes that she is the expert.

Selecting Courses to be Substituted

Once a student with a learning disability has been granted a mathematics course substitution, the next step is to decide on an appropriate course for the substitution. Courses to be substituted usually are recommended based on two different perspectives:

- relevant to mathematics
- relevant to the student's major.

However, the decision to suggest a course to be substituted has to be based on the student's processing disorders and processing strengths.

A student with an abstract reasoning processing disability should not have a chemistry course offered as a substitute (this is not a far-fetched ... it actually happened and the student failed the course!). Going back to Chapter Two, you need to know the effects your student's processing disorders on learning. Otherwise, you may inadvertently suggest a course substitution that the student cannot pass. Course substitutions that have the student use the same dysfunctional processing disorder are inappropriate and discriminatory.

Some postsecondary institutions want their students to substitute courses that are similar to mathematics. The student's the last passed math course can be used as the substitution for the remaining math course(s). Then, electives courses can be used for any remaining credit hours needed for graduation. If the institution suggests that the remaining courses must have a mathematics base, then the following courses are suggested, depending on the students processing disorders:

- Philosophy of Logic
- Business Mathematics
- Accounting
- Introduction to Computer Literacy
- Critical Thinking
- Astronomy
- Personal Finance

The Philosophy of Logic course is probably the most frequent substitution. Logic is the basis of mathematics. This is a good substitution, if the course only requires verbal reasoning skills, instead of abstract reasoning skills. Introduction to Computer Literacy is another popular course substitution. It requires minimum programming skills and is an overall benefit to the student. However, accounting may not be a good course substitution for students who have visual processing disorders and abstract processing disorders. The visual processing disorder will cause major problems in doing debit and credit ledgers, even if the students understands the abstract concepts. The Disability Support Service Provider must suggest substitution courses that can benefit the student and that the student can pass.

Course substitutions relevant to the student's major are easier to select. These courses should be selected with input from the student's academic advisor and the student's Disability Support Service Provider. English and Journalism majors usually select Introduction to Computer Literacy courses or Word Processing courses which can enhance their writing skills and are needed for future employment. To better understand their students, education majors, may select additional psychology courses. Art majors may select Humanities or History courses to better understand the relationship between art and the world. History majors may select Anthropology courses to understand how man's evolution effected history. Students can also select course substitutions in their same field of study. Selecting courses relevant to the student's major should expand the student's knowledge in the area and/or better prepare him/her for employment.

If your postsecondary institution does not have a course substitution policy procedure, it is the responsibility of the Disability Support Service Provider to develop one for the administration. This also is true if your current substitution procedure is vague or biased. Remember, the Disability Support Service providers are the best qualified persons to develop this procedure.

The discussed substitution procedures are examples that can be used as guidelines. However, for additional input, you should check with your sister institutions for their course substitution procedures. Each postsecondary institution needs to develop a course substitution policy/procedure, and it must be consistently followed. Having a publicized and fair course substitution procedures will benefit the institution, Disabled Student Services providers, and most of all the student with LD.

Chapter 6

Student Case Studies

6

This section is about students with LD who have applied for mathematics course substitutions. These cases are from different institutions with different course substitutions polices. Each case will be discussed in detail and the reasons for course substitutions or denial of course substitutions are given. As you read each case, look for similar students with whom you are working to understand how the process can be applied.

Student A

Student A is a 22-year-old woman, referred to the Disabled Students Office due to three failures in mathematics courses. She is an education major and has already taught students. Her advisor wonders if there are reasons for the mathematics course failures. The student is ready to graduate and is looking for employment. She is requesting a mathematics substitution but has not been diagnosed as LD.

The **WAIS-R** indicated a full scale IQ of 86. The verbal IQ score was 90 and the performance IQ was 83. She had low average intelligence. The scores from the **WJ-R Test of Cognitive Ability** indicated a severe long-term retrieval learning problem and a possible fluid reasoning learning problem. Even though she had average intelligence, her long-term retrieval standard score was 61 (kindergarten level) or a 0.5 percentile rank score. The difference between her IQ and long-term retrieval scores was 25 points. The long-term retrieval tests measured her ability to remember information over a period of about five minutes when learning was interrupted. This means that over 99 percent of the students had a better long-term retrieval skills than she did. Her fluid reasoning standard score was 80 (fourth grade level) or a 10 percentile rank score. This means that 90 percent of college students had better fluid reasoning skills than she did. Her short-term memory and other processing standard scores were average and above her IQ score. Her mathematics aptitude standard score was 86 and her broad math (mathematics achievement) standard score was 88. There was not a significant difference between her mathematics achievement score and her mathematics aptitude or intelligence score.

An intra-cognitive discrepancies assessment of the processing disorders was conducted to determine whether there was a significant difference between the processing disorders. Student A had a 2.36 standard deviation difference between her long-term retrieval score and the other processing scores. She had a 29 point difference between her expected long-term retrieval and her actual score which only happens in 1% of the population. On the other hand, her comprehension-knowledge processing score was 1.44 standard deviations above the other scores. Her high comprehension-knowledge score (12.4 grade level, 98 SS) probably accounts for her ability to learn other course material. However her other processing deficits blocked her ability to learn mathematics.

Student A was learning disabled. Her long-term retrieval prevented her from accurately keeping mathematics steps in her mind long enough to understand the concept. This problem manifested itself in not being able to build mathematics concepts that are required for future leaning. However, she had not received any learning or testing accommodation which could have decreased the effects of her learning disability. Wanting her to have a chance to learn as much mathematics as possible, the course substitution was denied. The student was instructed to take the mathematics course again with appropriate learning accommodation, testing accommodation and study skills training. The course substitution would be granted the next semester, if she tried to pass the course, using the recommended suggestions but failed it.

This type of situation happens more often than you might believe and can prevent students from graduating on time. Many students with LD leave high school, not being diagnosed, and have difficulty learning college level algebra. These students usually fail an algebra course or two, then wait until they are ready to graduate to attempt the algebra course again. It is difficult to locate these students, but working closely with the mathematics faculty is a good start. Be alert for this pattern even with your own students with learning disabilities.

Student B

Student B is a twenty-year-old male who was diagnosed as LD in elementary school. In college, he had math study skills training but was still having difficulty passing his basic algebra course. He has used learning accommodations and testing accommodations and tutorial service. After failing his third attempt in a non-credit algebra course, he applied for a course substitution. He was an English major with a 3.3 GPA and only needed his mathematics courses to graduate from the junior college.

As determined by the **WAIS-R**, Student B had a full Scale IQ of 91. His verbal IQ was 96 and his performance IQ was 87. His intelligence score was in the average range. The **WAIS-R Arithmetic and Vocabulary** subtest scores showed a significance weakness. The **WJ-R Test of Cognitive Ability** indicated his visual processing speed test score was low. The visual processing speed standard score was 83 which is at the seventh grade level. This score translate to the 13th percentile in visual processing speed. However, his major processing problem was in fluid reasoning where he scored an 83 standard score, which placed him at the fifth grade level. The fluid reasoning relative mastery index is 63/90. This means that while an average student was understanding 90 percent of the abstract mathematics material, he was only understanding 63 percent.

Looking at Student B's past history revealed additional information. Student B had attended a high school for students with learning disabilities. He demonstrated excellent language skills but had difficulty learning mathematics. Even with special instruction and tutoring Student B's mathematics achievement did not improve over a three-year-period from grades 10 to 12. Student B had good arithmetic skills but could not understand algebra. This was proven again when Student B passed his college arithmetic course but still could not pass the required algebra courses.

It is apparent that Student B's fluid reasoning problem has caused the mathematics learning problem and even with accommodations, he had maximized his mathematics learning. In fact, he probably learned as much mathematics as possible in high school. This math substitution should be granted. The only real question is why it took so long for Student B to apply for the substitution. To answer this question, you might have to understand the institutions substitution philosophy.

Student C

Student C is a twenty-eight-year old male who is majoring in physical education. Student C was diagnosed as having a learning disability after being referred to the disabled student office for mathematics learning problems. He made an "A" in the math study skill course but still continued to have difficulty learning basic arithmetic. Student C received learning and testing accommodations along with tutorial services, while attempting his second arithmetic course. He has a 2.4 GPA with 49 hours and wants to transfer to the university. He is requesting a mathematics substitution, after failing his third arithmetic course.

Student C has scored a 98 on his IQ test which demonstrates average intelligence. The scores from the **WJ-R Test of Cognitive Ability** indicated a severe auditory processing and fluid reasoning disorder. The fluid reasoning standard score was 77 which is at the fourth grade level and a six percentile rank. The fluid reasoning relative mastery index is 50/90. This score indicates that while an average student is understanding 90 percent of the abstract learning in a mathematics class Student C is only understanding 50 percent of the material.

His auditory processing standard score was a 75 which is at the second grade level, representing a five percentile rank. This means that 95 percent of the students were auditorial processing the lecture information better than student C. This also means that his intelligence score may be higher than indicated because his IQ test was verbally administered. This low test score definitely indicates that Student C was having difficulty understanding some of the mathematics lecture. Learning accommodations solved most of his auditory processing problem but could not solve all his fluid reasoning problem.

Student C was having difficulty learning mathematics due to his abstract reasoning disorder. He had shown effort in trying to learn the mathematics material and used the appropriate accommodations. Since he used a calculator to do the arithmetic functions required for a physical education major, his course substitution was be granted.

Student D

Student D is a woman in her mid forties who had been trying to complete her mathematics requirement to graduate for over three years. She had two unsuccessful attempts at basic algebra then took a basic arithmetic course and withdrew due to a failing grade. She had two more unsuccessful attempts at basic algebra before requesting a substitution. She also reported having difficulties learning mathematics through out her schooling. Student D had a 3.5 college GPA with 51 completed hours towards her Associate of Arts degree. Her major is not related to mathematics.

Student D has high average intelligence based on the **Slosson Intelligence Test.** Her IQ is a 113. Using a 95 percent confidence level her IQ is between 104 and 122. The percentile rank of her IQ is 79.

Student D's **WJ-R** test scores were compared to the intelligence score of 113. Student D's short-term memory standard score was 89 which is at the seventh grade level or a 23 percentile rank score. Her visual processing speed standard score was 72 which is at the sixth grade level or a three percentile rank score. The comprehension-knowledge standard score was 124, suggesting a senior college level or a 95 percentile rank score. This high comprehension-knowledge score indicates that she has excellent ability to learn general material. The long-term retrieval standard score is 85 which is at the third grade level or a 16 percentile rank. However, it is her fluid reasoning processing that is causing most of the mathematics learning problems. Her fluid reasoning standard score is 78 which is at the fourth grade level indicating a seven percentile rank. Her fluid reasoning relative mastery index is 53/90 meaning that she is understands only 53 percent of abstract mathematics concepts when the average student is understanding 90 percent.

An intra-cognitive discrepancies assessment of the processing disorders was conducted to determine if there was a significant difference between the processing disorders. Student D had a 1.27 standard deviation difference between her fluid reasoning score and the other processing scores. She had a 15 point difference between her expected fluid reasoning and her actual score which only happens in 10 percent of the population. On the other hand, her comprehension-knowledge processing score was 3.33 standard deviations above the other scores. Her high comprehension-knowledge score probably accounts for her ability to learn other course material. However, her fluid reasoning and long-term retrieval processing deficits blocked her ability to learn mathematics.

Her mathematics aptitude standard score was a 83 or 13 percentile which also indicated a mathematics learning problem. This score means the 87 percent of college students have a better mathematics learning potential than she does. Student D's broad math achievement standard score was 98 which is at the 13th grade level or a 45 percentile rank score. Student D's actual mathematics achievement score was .8 standard deviations above her expected mathematics achievement score. The mathematics aptitude/achievement discrepancy suggest that Student D has learned a little more mathematics than expected.

It is apparent that Student D's fluid reasoning problem is the major caused the mathematics learning problem along with her long-term retrieval problems. Her accommodations and effort in attempting five mathematics courses has enhanced her mathematics achievement about one standard deviation above her predicted level. This is probably all the mathematics she is capable of learning. The mathematics substitution request was approved because of the stated reasons and because her major does not require algebra.

Student E

Student E is a forty year old male who is majoring in criminal justice. In the fifth grade he was identified as learning disabled. In college he successfully completed an arithmetic course but did not pass a non-credit algebra course in four attempts. He has a 2.75 GPA with 36 earned credits. He is requesting a substitution for his remaining mathematics courses.

The **WAIS-R** results indicated that Student E had a Full Scale IQ of 118. His Verbal IQ was 114 and his Performance IQ was 117. His intelligence is in the above average range. His IQ is at the 88 percentile rank. His Digit Span Subtest, which measures short-term auditory memory and concentration, was a significant weakness. He had a two percentile Digit Span subtest percentile subtest rank. His Similarities Subtest score was a significant strength which measures verbal concept formation, non-verbal reasoning, and higher level cognitive skills. He also had a significant strength in Object Assembly subtest which measures visual analysis and perceptual organization skills. His second significant weakness was identified in the Digit Symbol subtest. The low Digit Symbol subtest score suggest poor or slow eye-hand coordination or deficits in immediate visual memory and rote learning of symbolic visual information. The **Visual Aural Digit Span Test (VADS)** result supported the findings of the **WAIS-R.** The **VADS** indicated that he was significantly below grade level on auditory memory and auditory writing tests.

The information from the **WAIS-R** and the **VADS** indicated that Student E has above average intelligence but has processing disorders. The major processing disorders are short-term auditory memory and concentration. The second processing disorder was is in rapid processing of visual information and visual memory. It does not appear the he has a long-term memory or abstract reasoning problem. However, it cannot be determined if Student E has a long-term retrieval or fluid reasoning problem with out **WJ-R** cognitive scores. The mathematics course substitution was approved due to Student E's effort (four math course attempts) and his ability to succeed in non-math related courses.

Student F

Student F is 20 year old male student who is majoring in film making. He was identified as learning disabled in seventh grade and was in LD classes in high school. Based on college placement tests, he was placed in a basic algebra course in his first semester. He withdrew from that course because he was failing. He had a 2.7 GPA with 22 hours when he requested a math course substitution.

As determined by the **WAIS-R** Student F had a Full Scale IQ of 117. His Verbal IQ was 111 and his Performance IQ was 125. His intelligence score was in the above average range. The significantly greater Performance IQ over the Verbal IQ indicates a pattern of better perceptual-motor skills than language ability or auditory processing. His Arithmetic subtest scale score was 10 (50 percentile). His Picture Arrangement and Block Design subtest scores were 13 (84 percentile) respectively. The Similarities scale score was 91 percent. The only significant subtest score difference was a strength in Object Assembly.

Student F's other documentation include report summaries of other tests but did not include **WJ-R** processing scores. From the reports, Student F has auditory and visual memory processing problems that effects his spelling and handwriting. The **WJ-R** cluster areas suggest adequate perceptual speed but poorly developed memory. His mathematics calculations are adequate but he has difficulty in application of complex equations. He has difficulty reading graphs, working with decimals, and using the metric system. He is functioning at grade level in his academics except for spelling.

The documentation presented suggest that Student F has a language disability especially in spelling probably due to his memory processing disorder. His **WAIS-R** subtest scale scores do not indicated any problems with abstract reasoning or long-term memory. His average score on the Arithmetic subtest is not significantly low and is probably due to his auditory short-term memory problem. It appears that this student is capable of learning more mathematics with the appropriate learning and testing accommodations along with study skills training. The course substitution was denied.

Since the course substitution was denied, Student F has not enrolled in another mathematics course. He now has 58 total hours and is about ready to graduate with an A. A. degree. Do you see Student F's potential graduation problem?

Student G

Student G is a 21 year old female student who was identified as being learning disabled in elementary school. She has had one unsuccessful attempt at a basic algebra course. Her GPA is 3.00 with 24 course credits and she has a history of withdrawing from courses. She is majoring in English and requested a math substitution.

Based on the **WAIS-R,** Student G had a Full Scale IQ of 100. Her Verbal IQ was 107 and her Performance IQ was 91. Her intelligence score was average. The significant superiority of the Verbal IQ over the Performance IQ indicates a pattern of better language or auditory processing skills than fine motor or perceptual-motor abilities. Scores on the subtests requiring verbal abilities were generally higher than those subtests scores which assess perceptual organization and nonverbal problem solving. The low score on Block Design indicates deficient ability in visual analysis and nonverbal problem solving.

Student G's **WJ-R** test scores were compared to the intelligence score of 100. Student D's auditory processing standard score was 80 which is at the third grade level or a 10 percentile rank score. The comprehension-knowledge standard score was 103, suggesting a freshman college level or a 57 percentile rank score. This comprehension-knowledge score indicates that she has the ability to learn general material. The long-term retrieval standard score was 84 which is at the third grade level or a 14 percentile rank. Her fluid reasoning standard score is 93 which is at the ninth grade level and at the 32 percentile. Her fluid reasoning relative mastery index was 85/90 meaning that she understands 85 percent of abstract mathematics concepts when the average student is understanding 90 percent.

The **WJ-R** mathematics applied problems test is another way to measure mathematics abstract reasoning. Student G's mathematics applied problems standard score was 77 which is at the sixth grade level representing a six percentile. The applied problems test's relative mastery index is 37/90. This score indicates that while an average student is understanding 90 percent of the abstract learning in mathematics class Student G is only understanding 37 percent of the material.

Students G's mathematics aptitude standard score was a 99 or 46 percentile. Her broad math achievement standard score was 74 which is at the sixth grade level or at the fourth percentile. The broad math achievement standard score consists of the applied problems score and calculation score. The calculation standard score was 72 which is at the sixth grade level or at the third percentile. Student G's mathematics aptitude/achievement discrepancy was -2.34 SD. This difference in scores happens in one out of a 100 students indicating a sever mathematics learning problem. Student G has a mathematics learning disability in both calculations and abstract reasoning.

An intra-cognitive discrepancies assessment of the processing disorders was conducted. The only cognitive discrepancy was a -1.22 standard deviation difference between her auditory processing score and the other processing scores. However, there was a significant intra-achievement discrepancy between her broad mathematics achievement and the other achievement scores. She had a -2.34 standard deviation difference between her mathematics achievement scores and her achievement scores for reading, written language and broad knowledge. This difference only happens in one percent of the population. It is apparent that Student G is learning other course materials but is not learning mathematics.

This is an interesting case because the student does not appear to have a significant long-term retrieval or fluid reasoning processing disorder. However, the **WJ-R** achievement tests indicates a mathematics reasoning problem and so does the Block Design subtest of the **WAIS-R.** The **WJ-R's** aptitude/achievement discrepancy suggests that the student was not meeting her potential. However, the intra-achievement discrepancy indicates that she has not achieve mathematics knowledge commensurate with her other subject areas. Since Student G's has attempted one course with the appropriate accommodations and no further action appear to improve her chances for success in mathematics learning, the substitution was granted.

Student HS

Student HS is a high school student who was be accepted to a major university in the southeast. She was diagnosed as having a learning disability when she was six years old. She always had difficulty in learning mathematics. In high school she failed or made "D"s in several pre algebra and algebra courses. However, the grades in her other subjects were excellent. While taking her algebra courses Student HS received learning accommodations, testing accommodation and tutorial assistance. Student HS is requesting a mathematics course substitution without attempting any college mathematics courses.

This is a very interesting case considering it may be the first time a major university as ruled on such a request. Student HS had to prove that she reached her mathematics potential and due to her learning disability could not learn any more mathematics. Then, Student HS had to show that substituting mathematics would not fundamentally alter her program. However, the university could not present their own documentation because she was not a current student. The documents had to come from the student. Student HS presented 28 documents including an opinion letter that I wrote (Reference E). Some of the supporting documents were psycho-educational evaluations, tutor notes, transcripts, instructor letters, individual education plans, recommendations, and three opinion letters. The most important documents will be used to explain this case.

HS's psycho-educational evaluations and school records indicated that she has a mathematics learning disability. According to the psycho-educational evaluations, Student HS's mathematics problems areas are, "... in applying estimation concepts and skills to whole and rational numbers and computations. Student HS also demonstrated below average performance in understanding and solving non-routine problems." Her high school records indicate an above average achievement in non-algebra related mathematics courses, but a below average achievement and failure in algebra related courses.

Based on Student HS's psycho-educational evaluations she has a full scale intelligence score of 118 indicating an intelligence range between 115 and 121. Her intelligence is considered to be in the high average bright and superior area. However, her intelligence subtest scores indicate a significant superiority of the Verbal score over the Performance score. The significant superiority of the Verbal over the Performance scale indicates a pattern of better language or auditory processing than fine motor or perceptual-motor abilities. This intelligence difference also suggests using the Verbal intelligence score as a more valid indication of her intelligence. Using her Verbal intelligence score of 122 indicated an intelligence range from 117 to 127. Her intelligence score of 122 places her in the Superior area.

According to the psycho-educational evaluations, Student HS demonstrates processing disorders in areas of auditory short-term memory, auditory attention span, reproductive visual-motor coordination and abstract visual analysis and synthesis. According to Student HS's computer generated intelligence report, there is a significant difference of - 4.2 in the Block Design Subtest. This demonstrates a significant weakness in concrete and abstract conceptualization, along with poor visual analysis. A weakness in Block Design also indicates a non-verbal abstract reasoning problem which is highly correlated to mathematics success. In contrast, Student HS has a significant strength in the Similarities verbal intelligence subtest indicating superior skills in verbal concept formation and verbal reasoning. In summary, excellent verbal reasoning skills are over shadowing her deficient abstract reasoning and auditory processing weaknesses. This might lead some professionals to overestimate her ability to complete college level algebra courses.

With an IQ of 122 Student HS should be able to successfully solve mathematics calculations and understand algebra concepts. Student HS has been tutored for years in mathematics and has attempted several pre algebra and Algebra I courses. Based on her instructor reports, tutor notes, and achievement test, and grades, Student HS has basic calculation shills but cannot solve algebra equations. Even with accommodations she could not comprehend or generalize the algebra concepts required to solve algebra equations.

Student HS wrote a letter to the Committee on Student Learning Disabilities regarding the effects of course substitutions as a fundamental alteration of her program. Student HS's program was defined as her major, political communications, instead of a B.S. degree. A political communications major does not require high level algebra skills to be successful in the required course work. Using a calculator can solve her mathematics problems while in the work place. Since her course work does not require algebra skill beyond her level, a mathematics course substitution would not fundamentally alter her program.

Another opinion letter was written by a clinical psychologist on the psychological effects of taking a mathematics course knowing that it would be failed. The psychologist opinion was that remediation is sensible as an initial trial to help the student reach her capability. However, continuing to endure the ongoing negative emotions associated with failure, when maximum achievement has be met, leads to negative self-esteem. This is especially true when having evidence of a correlation between learning disabilities and emotional difficulties in adulthood. In this case, a course substitution would minimize the risk of emotional problems. The psychologist concludes his opinion with this statement, "The bottom line is that if (Student HS's) mental health is jeopardized, all of course work and other accomplishments may be irrelevant." From the psychologist opinion, it appears that taking another mathematics course just to prove that she cannot do algebra is psychologically unsound.

Student HS's documentation proved that her mathematics potential was reached and her learning disability blocks her ability to learning additional mathematics. Documentation was presented on the detrimental psychological effects of just trying algebra one more time. Student HS also showed that substituting mathematics would not fundamentally alter her program. She also received a letter of substitution recommendation from the disabled student coordinator.

As you have already guessed, Student HS was granted a mathematics course substitution. The courses substituted were PHI 2100 - Reasoning and Critical Thinking and CLT 3041 - Word Building: Greek and Latin Elements in the English Vocabulary. It appears that the PHI 2000 course is in the mathematics area and the CLT 3041 is in the communications area.

Conclusion

Students with learning disabilities are continuing to increase in numbers at postsecondary institutions. Some of these students due to the nature of their learning disability cannot pass required mathematics courses. Section 504 allows for course substitutions when the course substituted does not fundamentally alter the student's program. Postsecondary institutions must develop course substitutions procedures that are accessible for students with disabilities. Disability Support Service providers are the most qualified individuals to assist the institution in revising or developing a course substitution procedure. Disability Student Service providers have two different roles in revising or developing course substitution procedures.

The first role is educating the administrators/faculty about learning disabilities and mathematics, assisting students through the course substitution process, and writing a substitution recommendation letter. There are many different steps in accomplishing this role. Reference F (Mathematics Course Substitution Proposal Check List) was developed for Disability Support Service providers to organize these steps. Disability Support Service providers only have to check off the items that apply to their institution. This is not an all inclusive list of suggestions but should be used as an guideline.

Disability Support Service providers may need additional information or training on educating administrators/faculty about learning disabilities. The Association of Higher Education and Disability (AHEAD) has publications pertaining to students with LD in higher education. A list and description of these publications is in Reference G. I would alsop suggest that you join this excellent organization.

The second role of Disability Support Service providers is assisting the institution in developing or revising a course substitution procedure. This procedure should include all disabilities groups and for any type of course substitution. For example, the substitution procedure should be broad enough to include students with physical disabilities who are applying for English, foreign language, or physical education course substitutions. Reference H (Course Substitution Procedure Development/Revisement Check List) was developed to assist the Disability Support Service provider in accomplishing this goal. The check list is a guideline and you do not have to complete each item on the list. Review each item for its use at your institution. Talk with your supervisor and staff about the items you wish to accomplish. If you need legal information for developing or improving your substitution procedure then contact Heywood, Lawton, and Associates who are the authors of **Disability Accommodation Digest** fot their publications (Reference I). Remember, your institution must have an effective course substitution procedure.

In Chapter One it was stated that postsecondary institution administrators and Disability Support Service providers are struggling with three basic questions

- which students with LD can pass their required mathematics courses
- which students with LD do we grant course substitutions to
- which courses are appropriate substitutions?

The information in this book can be used as a guide to answering these three questions.

It appears that students with learning disabilities which involve problems with imputing information have a better chance of passing required algebra courses. Some examples of these disorders are short-term memory, processing speed, auditory processing, and visual processing. These students with appropriate study skills training, learning accommodations, and testing accommodations have a good chance of passing their required mathematics courses. However, since no assessment is a perfect measurement of a student's processing disorders, some of theses students may still not be able to pass required mathematics courses.

The research on students with severe processing disorders is clear: they have major mathematics learning problems. These severe processing disorders involve long-term memory, long-term retrieval, and fluid (abstract) reasoning. Theses student usually input mathematics information correctly but cannot remember the information or cannot understand the concept. This author has worked with many students with severe processing disorders and believes that students with fluid reasoning problems have the most difficulty learning mathematics. Students with fluid reasoning problems can only learn/generalize mathematics concepts to a certain level. And once they reach that level they cannot learn anymore mathematics no matter what we do, Students and instructors describe this phenomenon as "hitting the brick wall". Students with severe processing disorders need to learn as much mathematics as possible and then be allow to substitute the remaining mathematics course(s).

Once a student with LD has been granted a mathematics course substitution, the next step is to decide on an appropriate course substitution. Courses to be substituted usually are recommended based on two different perspectives

- relevant to mathematics
- relevant to the student's major

The courses that were suggested for substitution at different institutions or in the individual case studies were relevant to mathematics and the student's major. In some cases a course was selected from each area. The Philosophy of Logic course is probably the most frequent substitution for the mathematics area. This is a good substitution only if the course requires verbal reasoning skills, instead of abstract reasoning skills. Selecting a course substitution relevant to the student's major should include input from the Disability Support Service Provider, the student's academic advisor, and the student. **Remember, the course substitution decision must be based on the student's processing disorders and processing strengths. Approving course substitutions requiring the student to use his/her dysfunctional processing disorder(s) are inappropriate and can be discriminatory.**

References

Reference A

Learning Accommodations Fact Sheet

Problem 1

decimals to % move the decimal point two places to right and add a zero as needed and add the %

.36 = 36.%
36%

.457 = 45.7%
45.7%

.2 = 20.% add 0
20%

fraction to % enter numerator in your calculator and divide by the denominator. This makes a decimal and now do the same as above.

$\frac{1}{4} = .25 = 25\%$

Percent Problems

60 is 30% of what number?

is means = , of means multiply, what number =

60 = (30%)n change % to fraction and solve

$60 = \frac{30}{100}n$ divide by $\frac{30}{100}$ (both sides)

$$\frac{60}{\frac{30}{100}} = \frac{\frac{30}{100}}{\frac{30}{100}}n$$

n = 200

Problem 2

$$\frac{8.5 \times 10^{8}}{3.4 \times 10^{-5}}$$

[8.5] [x] [10] [INV] [8] [=] (display shows 8.5^{08})

[÷] [3.4] [÷] [10] [INV] [5] [±] [=]

display: 2.5^{13}

means: 2.5×10^{13}

Problem 3

$(1.123 \times 10^{4}) \times 10^{-9}$

[1.123] [x] [10] [INV] [x] [4] [x] [10] [INV] [9] [±] [=]

display 0.0000112 (1.12×10^{-5})

moved decimal point to the *right* five places so the exponent becomes -5

display: 1.123^{-14}

means: 1.123×10^{-14}

Problem 4

$(5.2 \times 10^{5})(6.5 \times 10^{-2})$

[5.2] [x] [10] [INV] [x] [5] [=]

base number — base — exponent

[x] [6.5] [x] [10] [INV] [x] [2] [±] [=]

[33800]

Want the decimal point after the first non-zero number. Drop trailing zeroes.

33800

3.38×10^{4}

moved the decimal point to the *left* four places so the exponent became a +4

Problem 5

$(x2 - 10x - 25) \div (x - 5)$

two terms so have to use *long* division

```
         x      -5
x-5 | x2    -10x    -25
      x2  (+)-5x
      0     -5x     -25
         (+)-5x (-)+25
             0      -50
```

change signs
and add

can't put variable x
into constant -50
so that is a remainder

$$x - 5 - \frac{50}{x-5}$$

Problem 6

```
    3x^4
8 | 24x^4  -4x^3  +x^2  -16
   -24x^4
       0  -4x^3  +x^2  -16
```

do this way when only one number or variable is in the denominator

$$\frac{24x^4 - 4x^3 + x^2 - 16}{8} = \frac{24x^4}{8} - \frac{4x^3}{8} + \frac{x^2}{8} - \frac{16}{8}$$

$$= 3x^4 - \frac{1}{2}x^3 + \frac{1}{8}x^2 - 2$$

Problem 7

Percents *% to fraction* % means part per hundred so remove the % sign and make a fraction with *100 as the denominator.* Put the fraction in the calculator and hit the = sign to reduce to the lowest terms.

Examples: $70\% = \frac{70}{100} = \frac{7}{10}$

$63\% = \frac{63}{100} = \frac{63}{100}$

$125\% = \frac{125}{100} = 1\frac{1}{4}$

$.2\% = \frac{.2}{100}$ display .22100 $= \frac{1}{500}$

.2% = .2/100 Display = 1/500

% to decimals: drop the percent sign and move the decimal point two places to the left. If there was no point assume there is to the right 50 = 50. and you may have to add 0's as needed on the left.

40% = 40.% = .40
.5% = .005 here you have to add two zeroes as place holders
56.3% = .563

Reference B

Final Exam Fact Sheet
Order of Operations

P E MD AS

Work from left to right and do the inner most *( ) first.* Use your calculator.

Exponents: Anything to the O power = 1 no matter how long or if there are other exponents inside the parenthesis.

Examples:

$(2x \quad + 3y \quad - ab\ ) = 1$

$(14x\ y\ z\ a\ b\ c\ ) = 1$

A negative sign with exponents

$-3 \quad = -9$

$(-3) = 9$

$-(-3) = -9$

Negative Exponents

$a^{-1} = \frac{1}{a}$

$b^{-2} = \frac{1}{b^2}$

$\frac{3}{a^{-2}} = 3a^2$

$\frac{1}{3^{-1}} = 3$

either make a fraction or bring the negative exponent to the top and have a whole number/variable

when evaluating with negative exponents first make the exponent positive and then put the number in place

Example:

Evaluate: a when a = 3

$\frac{1}{a^2} = \frac{1}{3^2} = \frac{1}{9}$

Evaluate:

$$\frac{a^{-4}}{b^{-3}} \quad \text{when } b = 2,\ a = 3$$

$$\frac{a^4}{b^3} = \frac{3^4}{2^3} = \frac{81}{8}$$

Evaluate:

$$\frac{x^{-6}y^7}{z^{-2}} \quad \text{when } x = 2,\ y = 3,\ z = 4$$

rewrite with positive exponents *first*

$$\frac{z^2 y^7}{x^6}$$

then "plug" in the numbers and *show the next step*

$$\frac{4^2 3^7}{2^6}$$

then use your calculator

Solving Equations: Do *one step at a time* even though it takes longer. You are not in a race. There are several different kinds of equations aso there is no general format except to isolate the variable on one side using addition, subtraction, multiplication, and division as necessary. Use your calculator whenever possible.

Examples:

a: $\frac{1}{2}x = 6$

$$\frac{\frac{1}{2}x}{\frac{1}{2}} = \frac{6}{\frac{1}{2}} \quad \text{divide both sides by } \frac{1}{2}$$

$$x = 12$$

b: $-\frac{1}{3}y = 4$

$$\frac{-\frac{1}{3}y}{-\frac{1}{3}} = \frac{4}{-\frac{1}{3}} \quad \text{divide both sides by } -\frac{1}{3}$$

$$y = -12$$

Reference C

Interpretation of a State Substitution Procedure

Draft Copy

Substitute Admission and Graduation Requirements
for Handicapped Students

Policy 6Hx28:5-14

General Authority: 240.319 FS

Law Implemented: Section 240.152
and 240.153, FS. and Rule 6H-1.041, SBCC

Policy:

Admission to the College

Any person who is hearing impaired, visually impaired, or dyslexic, or who has a specific learning disability, shall be eligible for a reasonable substitution for any requirement for admission to the college where documentation can be provided that the person's failure to meet the admission requirement is related to the disability. The president or his designee is authorized to develop substitute admission requirements for each such person as shall be appropriate.

Admission to a Program of Study and Graduation

Any student who is hearing impaired, visually impaired, or dyslexic, or who has a specific learning disability, shall be eligible for reasonable substitution for any requirement for graduation, or for admission into a program of study where documentation can be provided that the person's failure to meet the requirement is related to the disability, and where the failure to meet the graduation requirement or program admission requirement does not constitute a fundamental alteration in the nature of the program. Upon receipt of documentation which substantiates that the disability can be reasonably expected to prevent the student from meeting the requirements for admission to a program of study or graduation, the president or his designee is authorized to develop substitute admission or graduation requirements for each such person as shall be appropriate for the particular program or course of study for which the student seeks admission or is enrolled.

Right of Appeal

Any person seeking substitution under this policy who is aggrieved by a decision of the president or his designee in the application of this policy shall have the right to appeal such decision under Policy 6Hx28:10-15, procedure B, as an appeal of an administrative decision.

Procedures:

The provost of the campus at which the application for admission is pending or at which the student records are maintained shall be designated as the person to make the determination of substitute admission and graduation requirements.

In determining whether to grant a substitution, documentation to substantiate that the disability can be reasonably expected to prevent the individual from meeting requirements for admission to the institution, admission to a program of study, or graduation shall be provided by the student as requested by the college.

For purposes of implementing this policy, the following definitions shall apply:

(a) Hearing Impairment. A hearing loss of thirty (30) decibel or greater, pure tone average of 500, 1000, 1000 Hz, ANSI, unaided, in the better ear. Examples include, but are not limited to, conductive hearing impairment or deafness, sen sorineural hearing impairment or deafness, high or low tone hearing loss or deafness, and acoustic trauma hearing loss or deafness.

(b) Visual Impairment. Disorders in the structure and function of the eye as manifested by at least one of the following: visual acuity of 20/70 or less in the better eye after the best possible correction, peripheral field so constructed that it affects one's ability to function in an educational setting, or a progressive loss of vision which may affect one's ability to function in an educational setting. Examples include, but are not limited to, cataracts, glaucoma, nystagmus, retinal detachment, retinitis pigmentosa, and strabismus.

(c) Specific Learning Disability. A disorder in one or more of the basic psychological or neurological processes involved in understanding or in using spoken or written language. Disorders may be manifested in listening, thinking, reading, writing, spelling, or performing arithmetic calculations. Examples include dyslexia, dysgraphia, dysphasia, dyscalculia, and other specific learning disabilities in the basic psychological or neurological process. Such disorders do not include learning problems which are due primarily to visual, hearing or motor handicaps, to mental retardation, to emotional disturbance, or to an environmental deprivation.

Upon the request of an applicant for admission or a student on a form provided by the college for substitution due to disability, the provost shall investigate, or cause an investigation to be made, to substantiate the documentation and to identify reasonable substitutions as to criteria for admission to the institution, admission to a program to study, or graduation as it relates to each such disability on an individual basis. Upon conclusion of the investigation and upon making such determination as is appropriate, the provost shall notify the student in writing of such determination. The determination shall become a part of the permanent record of the student and be forwarded to other state institutions upon the request of the student. A student or person seeking admission under this policy may appeal a denial of a substitution or determination of ineligibility as determined by the provost to the committee under the provisions of Policy 6Hx 28:10-15, Procedure B, as an appeal of an administrative decision.

Reference D

Course Substitution Request Correspondence

MEMORANDUM

TO: Assistant Dean
FROM: Disabled Student Services
DATE:
SUBJECT: SS# - -

Substitutions for General Education Program course requiring algebra and the Gordon Rule mathematics requirement has been diagnosed as having a specific learning disability related to which makes learning higher level/abstract mathematics extremely difficult (see attached documentation). Under BOR Rule 6C-6.01B Substitution of Course Required for Program Admission and for Graduation by Handicapped Students, Mr/Ms. ____ is requesting the following General Education Program (GEP) substitutions of courses:

1. One course substitution for the GEP Mathematical Foundations requirement and the Gordon Rule. ____ has completed CGS 1060 for 3 credit hours of the Mathematics Foundations requirement because this course does not require mathematics. Therefore, substitution of one course for the other 3 credit hours of the Mathematical Foundations requirement and the Gordon Rule is requested.

2. Two science courses from only he second section of the GEP Science Foundations requirement substituted for one science course from the first section and one science course from the second section. Mr/Ms.____ is unable to successfully complete one science course in the first section of the Science Foundations requirements because all these courses have a prerequisite of either MAC 1104 or MGF 1203, the mathematics course for which s/he is requesting a substitution.

If this request is approved, please include a list of courses which may be substituted to fulfill the GEP Mathematics Foundations requirements. Please inform Mr/Ms. ____ of your decision regarding this request:

student name
street address
city, FL xxxxx

Letter to Student

June, 5 1992

Student name
Address
City, FL xxxxxx

Dear Student:

Based on the information provided by the Director of Disabled Student Services regarding your difficulties with mathematics, I have made the following decisions.

1. You will be allowed to substitute MAT 0003 from ______ Community College for the University's Math Foundation section of the University's General Education Program.

2. You will be allowed to take two science courses from the second section of the Science Foundations part of the University General Education Program.

This approval does not affect any requirements which are called for by your major: any substitutions in that area must be approved by the department.

Best wishes for the successful completion of your program of study.

Sincerely,

Academic Dean

cc: Director of Disabled Student Services
Communications Dept.

Reference E

Student Substitution Request Letter

Paul Nolting, Ph.D.
Learning Specialist

July 2, 1992

Dr. W
Office of Disabled Students Services
University Y

Dear Dr. W:

Mr. X requested that I review his daughter's records and make a recommendation on her ability to complete college level mathematics courses based on her learning disability. I am a Learning Specialist at Manatee Community College in Bradenton, Florida and have given numerous workshops on learning disabilities, especially in the area of mathematics. I have worked with hundreds of students with learning disabilities and I am a author of books on this subject. (See enclosed vita and books).

From reviewing Students HS's Psycho-Educational evaluations and school records, it is clear that Student HS has a mathematics learning disability which the state of Florida describes as dyscalculia. According to the Psycho-Educational evaluations, Student HS's mathematics problems areas are, "... in applying estimation concepts and skills to whole and rational numbers and computations. Student HS also demonstrated below average performance in understanding and solving non-routine problems." Her high school records indicate an above average achievement in non-algebra related mathematics courses, but a below average achievement and failure in algebra related courses. Student HS's 340 SAT math score indicates a lack of understanding of algebra which is verified by her score of 13 on her ACT math section. As you know, the state cut off score for the first college credit algebra course (MAT 1033) excluding Gordon rule courses offered by (MAC 1102, MAC 1104, ...) is a math score of 16 on the enhanced ACT and a 400 on the math part of the SAT. If Student HS were to attend Manatee Community College she would be placed in MAT 0002 which is an arithmetic course.

A summary of Student HS's mathematics achievement suggests average arithmetic skills, but she lacks the understanding to apply concepts, solve non-routine problems, and solve non- credit algebra course problems. This lack of achievement is not due to lack of exposure or effort.

The focus now shifts to determining the degree of her learning disability and deciding if it will continue to effect her mathematics learning. Based on Student HS's Psycho-Educational evaluations she has a full scale intelligence score of 118 indicating an intelligence range between 115 and 121 which is in the High Average Bright and Superior area. However, when I used a computer generated report, using her intelligence subtest scores there is a significant superiority of the Verbal score over the Performance score. This intelligence difference suggests using the Verbal intelligence score as a more valid indication of her intelligence. Using her Verbal intelligence score of 122 indicated an intelligence range from 117 to 127. Her intelligence score of 122 places her in the Superior area.

As indicated in Mr. X's letter to the University Y's Office of Disabled Students, Student HS could be in the gifted area due to research indicating that learning disabled individuals consistently score lower on intelligence test due to their specific processing disorder(s).

According to the Psycho-Educational evaluations, Student HS demonstrates processing disorders in areas of auditory short-term memory, auditory attention span, reproductive visual-motor coordination and abstract visual analysis and synthesis. According to Student HS's computer generated intelligence report, there is a significant difference of - 4.2 in Block Design subtest. This weakness demonstrates a significant weakness in concrete and abstract conceptualization. A weakness in Block Design also suggests a non-verbal abstract reasoning problem which is highly correlated to mathematics success. In contrast, Student HS has a significant strength in the Similarities verbal intelligence subtest indicating superior skills in verbal concept formation and verbal reasoning. In summary, HS's excellent verbal reasoning is over shadowing her deficient abstract reasoning and auditory weaknesses. This might lead some professionals to overestimate her ability to complete college level algebra courses.

To help learning disabled students become more successful in mathematics, a concentrated effort is needed to focus on improving students math study skills, learning accommodations, and testing accommodations. Math study skills improvement can compensate for some of the student's processing disorders. Appropriate learning and testing accommodations can compensate for additional processing disorders. In some student cases even study skills training along with learning and testing accommodations will not be sufficient to help every learning disabled student pass enough mathematics courses to graduate.

To help students improve their math study skills they are referred to **Winning at Math: Your Guide to Learning Mathematics Though Successful Study Skills** which has study suggestion and a sperate section for learning disabled students. Students can also take the **Winning at Math: Study Skills Computer Evaluation Software** which analyzes their math study skills and gives specific suggestion to learning disabled and non-learning disabled students.

Math and the Learning Disabled Student: A Practical Guide For Accommodations suggests different learning and testing accommodations for students with different types of processing disorders. In most cases students with visual processing speed, visual processing, auditory processing, and short-term memory problems can be given appropriate accommodations to learn mathematics. However, students with abstract (fluid) reasoning learning disabilities will only be able to learn mathematics (algebra) to a certain level before requiring a substitution.

If Student HS has been taught math study skills and been given appropriate learning and testing accommodations then she probably has reached her maximum mathematics achievement level.

Student HS has auditory short-term memory, auditory attention, and visual processing disorders that probably have been or can be accommodated. This leaves the major problem of trying to accommodate Student HS's abstract reasoning learning disability. Abstract reasoning accommodations are too numerous to mention but can be reviewed in **Math and the Learning Disabled Student**. Even using all the abstract reasoning learning accommodations does not guarantee complete learning compensation or algebra course success.

The second major problem is placing Student HS into the correct mathematics course. According to Bloom (1976) fifty percent of a student's grade is based on cognitive entry skills and intelligence. This fifty percent of a students grade is gain or lost based on placement. Placing Student HS into an inappropriate mathematics course even without considering her learning disability suggests course failure.

Based on Student HS's test scores she needs to be placed into MAT 0002 which is a non-credit arithmetic course. By taking a second placement test she might move up to MAT 0024 which is a non-credit algebra course. According to state placement practices placing her in MAT 1033 should end in course failure and might be a violation of the state mandated placement requirements. Adding on the fact she has a mathematics learning disability and failed the second semester of algebra, I predict an even higher chance of MAT 1033 course failure. I am sure she would fail MAT 1033 even with appropriate accommodations.

Since University Y can not offer MAT 0002 or MAT 0024 and Z Community College does not offer these courses through University Y and for other states reasons, I am recommending a mathematics substitution for Student HS's Gordon Rule mathematics courses. If you have any questions about this report or other strategies to help your learning disabled students, call me at (813) 755-1511, ext. 4239 or SUMCOM 560-4239.

Sincerely,

Paul Nolting, Ph.D.
Learning Specialist

Reference F

Mathematics Course Substitution Proposal Check List

	Yes	No	NA*
1. Educated mathematics faculty about learning disabilities			
2. Educated administers about the effects of learning disabilities on mathematics learning			
3. Understand institutional's philosophy on mathematics course substitution			
4. Documented the student as being learning disability			
5. Documented the student as a "qualified disabled individual"			
6. Discussed course substitution procedures with student			
7. Obtained the student's substitution request letter			
8. Suggested that the student obtain a letter on the psychological effects of taking a math course			
9. Suggested that the student obtain substitution approval opinion letters regarding ability to learn mathematics			
10. Documented the processing disorder(s) which are causing the student's learning problem			
11. Documented the student's learning accommodations			

* *not applicable*

	Yes	No	NA*
12. Documented the student's testing accommodations			
13. Documented the student's mathematics study skills training			
14. Documented student effort in learning mathematics			
15. Obtained the student's academic record's sheet including GPA and number of completed hours			
16. Documented reasons for course withdrawals or poor grades			
17. Obtained the student's high school records and his/her Individual Education Plan			
18. Documented the student's declared major			
19. Documented that a mathematics course substitution will not fundamentally alter his/her program			
20. Considered other required courses that might have to be substituted due to math prerequisite			
21. Suggested appropriate course substitutions			
22. Obtained tutors' support letters			
23. Obtained mathematics instructor's support letter			
24. Complete the disabled student service provider's recommendation letter			
25. Discussed course substitution request with supervisor			

* *not applicable*

Reference G

Resources of Learning Disabilities in Higher Education

Association on Higher Education and Disability (AHEAD)
PO BOX 21192
Columbus, OH 43221-0192
614-488-4972 (V/T); 614-488-1194 (FAX)

Publications

Assisting College Students with Learning Disabilities: A **Tutor's Manual** This manual is designed for use by service providers and tutors working with students with learning disabilities. Included are suggestions for (1) determining problem areas, (2) helping students study effectively, (3) planning exam strategies, and (4) managing time. Also included is a sample tutoring program for spelling. By Pamela Adelman and Debbie Olufs. $26.00 (AHEAD members-$16.50) (1990-Second Printing)

Facilitating an Academic Support Group for Students with Learning Disabilities: A Manual for Professionals A product of a federal transition grant, this manual gives directions on how to develop and implement peer support groups for students with learning disabilities in higher education. It includes handouts, discussion group topics, facilitator instructions, and more. By Janis Johnson and Lynda Price. $23.00 (AHEAD members- $15.50) (1989-First Printing).

Journal of Postsecondary Education and Disability-Special LD Issue (Vol. 9, Nos. 1-2) Seven articles on service delivery for postsecondary students with learning disabilities. Topics include: transition issues, diagnosis, work success, social competence, and annotated bibliography. $20.00

Monograph on Multicultural Diversity and Learning Disabilities During the spring of 1990, more than 250 people spent more than 60 hours discussing the concerns that impact on culturally different postsecondary students with learning disabilities. The unique issues of Black students, Asian students, students of Spanish-language background, Pacific Islanders, Native Americans, and older, nontraditional students were discussed. This monograph is an attempt to summarize the wealth of information uncovered and begin to formulate both a plan of action an agenda for further exploration in determining how best to identify and serve the culturally diverse students who represent a growing segment of our service population. $23.00 (AHEAD members-$15.50) (1990-First Printing).

College Students with Learning Disabilities Brochures A general information brochure originally produced by the McBurney Resource Center. Current edition by Loring Brinckerhoff, Ph.D. $.35 (1-49 copies); $.30 (50-99 copies); $.25 (100+ copies)

Survival Kit for Learning Disabled Students in Higher Education (1988) A set of seven pamphlets by Jane Jarrow, Ph.D. and Loring Brinckerhoff, Ph.D., designed to orient the student with a learning disability to the issues and responsibilities of his/her participation in higher education. Camera-ready copy is provided for ease of reproduction. $32.95 (AHEAD members-$27.95)

Support Services for Students with Learning Disabilities in Higher Education: A Compendium of Reading, Book 3 The third volume of articles from AHEAD Publications that address service provision for students with learning disabilities in higher education. Topics include: definitions and diversity, legal issues, program development, assessment, strategies and techniques, social/emotional perspective, and transition. Edited by Mary Farrell, Ph.D. $40.00 (AHEAD members-$25.00)

Special Offer!

With purchase of Support Services for Students with Learning Disabilities in Higher Education ... Book 3 - Save and additional 33% when you purchase both the first and Second LD compendia! $48.40 (AHEAD members-$25.00)

Support Services For LD Students in Postsecondary Education: A Compendium of Readings (1987) Topics include: Transition from high school to college, model service delivery programs, academic accommodations, faculty awareness, psychosocial issues, and employment issues.
Regular price: $34.95 (AHEAD members-$15.50)

Support Services for LD Students in Postsecondary Education: A Compendium of Reading, Vol. 2 (1989) Topics repeat form the 1987 edition plus diagnostic guideposts, and computer access. Regular price: $37.95 (AHEAD members-$21.95)

AHEAD

An international, multicultural organization of professionals committed to full participation in higher education for persons with disabilities. The association is a vital resource, promoting excellence through education, communication, and training.

Membership Benefits:

- Annual Conferences
- Publication
- Membership Directory
- Employment Exchange
- Training
- Technical Assistance
- Information
- Professional Development
- Awards/Recognition
- Special Interest Groups
- Networking
- Discounts on publication, conferences, mailing labels

Membership Categories

Active Professional - $75.00 / Any person actively involved in working to enhance higher education opportunities to persons with disabilities. An active member is eligible to vote and hold office.

Affiliate - $65.00 / Any person supporting the purposes, goals, and objectives of the Association and choosing to make contributions in less visible or time-demanding roles. Affiliate members have voice, but may not vote or hold office.

Institutional - $187.50 / Any organization or institution. Each member institution is entitled to appoint one individual who shall be an Active Professional member, with all rights and privileges thereof.

Additional Professional - $37.50 / Any member organization or institution may appoint additional individuals to Active Professional membership in the Association at this reduced rate.

Student - $35.00 / Student membership is designed particularly for individuals whose primary involvement with higher education is that of being enrolled in graduate or undergraduate programs that prepare future disability service providers. Students members may vote and hold office.

Association on Higher Education and Disability
P.O. BOX 21192
Columbus, OH 43211-0192
614-488-4972 (V/T);614-488-1174(FAX)

Reference H

Course Subsitution Procedure Development/Revisement Check List

	Yes	No	NA*
1. Have you researched the course subsitution institutional prcedures?			
2. Have you researched recent outcomes of the course subsitution processes?			
3. Does the instutution have an operational defination of a learning disabilities?			
4. Is there a mechanism to identify students eligible to apply for course subsitutions?			
5. Is there a mechanism to identify reasonable subsitutions for graduation requirement for each disability?			
6. Is there a mechanism for making designed subsitutions known to the student?			
7. Is there a mechanism for making subsitution decisions on an individual basis?			
8. Is there a mechanism for students to appeal a subsitution denial?			
9. Is there an articulation provision with other institutions?			
10. Is there a mechanism to record the course subsitution to meet the student's graduation requirements?			
11. Is there a mechanism to maintain records on the number of granted subsitutions for each disability?			

* *not applicable*

	Yes	No	NA*
12. Is there a mechanism to maintain records on the number of denied subsitutions for each disability?			
13. Has the board of trustees approved the course subsitution procedure?			
14. Is the disabled student coordinator on the subsitution committee?			
15. Is there provisions for the LD specilist or learning specilist to be on the subsitution committee?			
16. Have the committee members been educated about the effects of learning disbilites?			
17. Have the committee members been educated on Section 504 and the ADA?			
18. Have the committee members been informed about the course subsitutuion procedure?			
19. Does the committe make suggestions for the courses to be subsituted?			
20. Have courses allready been selected for mathematics course subsitutes?			

* *not applicable*

Reference I

Resources on Legal Information

Heyward, Lawton, and Associates
PO Box 537
Narragansett, RI 02882
401-789-5089

Publications

Access to Education for the Disabled: A Guide to Compliance with Section 504 of the Rehabilitation Act of 1973 This book has pertinent parts of the 1973 education act and how to put them into effect are analyzed here. Hypothetical situations based on actual cases show how the act has been interpreted by the Department of Education's Office of Civil Rights. Partial contents: The Evolution of Compliance Standards...An Overview of Section 504...The Duty to Provide Access to Facilities...Elementary and Secondary Education...Postsecondary Education...Employment...Enforcement. This book is written by civil rights attorney Salome M. Heyward who is a nationally known educational consultant on disability discrimination issues. Call MaFarland and Company, Inc. Publisher, Box 611, Jefferson, NC 28640 at 919-246-4460 to order this book at $38.50.

10 Hours of Substantive Training in Section 504 and ADA Issues This new video training series by Salome M. Heyward, Esq. who has provided training products and services to over 250 educational institutions and state agencies has practical legal information and is up to date. Demo video are $15.00 and can be ordered through HL & A, Narragansett, RI 02882.

Disabilities Accommodation Digest This newsletter is published quarterly and contains information and articles of interest to persons concerned about the implementation and enforcement of Section 504 and the Americans with Disabilities Act. There are regular columns such as *Ask the Experts* and *Significant Cases and OCR Findings. On the Horizon* presents developing issues and trends as well as new products and services. $60.00/year

Additional Services

Equity Audits Heyward, Lawton, and Associates will conduct in-depth reviews of the pertinent policies and procedures of districts/institutions having difficulty complying with Section 504 and the Americans with Disability Act. Compliance problems will be identified and strategies will be developed to address them.

Complaints Assessment and Resolution Heyward, Lawton, and Associates staff will review complaints of disability/handicap discrimination and provide advice regarding the merits of the complaints and recommend possible solutions. They will also assist institutions in preparing for federal investigations.

Information Service Our staff will answer your questions regarding disability discrimination law. Our database and resources are available to you for on-the-spot answers to your compliance questions. We will also perform specialized research upon request.

Bibliography

Ackerman, P. T., Anhalt, J. M., & Dykman, R. A. (1986a). Arthmetic automatization faulure in children with attention and reading disorders. Associations and sequel. **Journal of Learning Disabilities,** 19, 222-232.

Algozzine, B. O'Shea, D.J., Crews, W. B.,& Stoddard, K. (1987). Analysis of mathematics competence of learning disabled adolescents. **The Journal of Special Education,** 21(2) 97-107.

Batchelor, E. S., Gray, J. W.,& Dean, R. S. (1990). Empirical testing of a cognitive model to account for neuropsychological functioning underlying arithmetic problem solving. **Journal of Learning Disabilities**, 23, 38- 42.

Beirne-Smith, M., & Decker, M. (1989). A survey of postsecondary programs for students with learning disabilities. **Journal of Learning Disabilities**, 22, 456-457.

Bigler, E. D. (1992). The Neurobiology and Neuropsychology of Adult Learning Disorders. **Journal of Learning Disabilities,** 25, 489-506.

Bireley, M. K. (1986). The Wright State University program: Implications of the first decade. **Reading, Writing, and Learning Disabilities**, 2, 349-357.

Bley, N. S., & Thornton, C. A. (1981). **Teaching Mathematics to the Learning Disabled**. Rockville, MD: Aspen System.

Blalock, J. W. (1980). Persistent auditory language deficit adults with learning disabilities. **Journal of Learning Disabilities**, 15, 604-609.

Brinkerhoff, L., Shaw, S., & McGuire, J.(1993). **Promoting Postsecondary Education for Students with Learning Disabilities.** Austin, TX: PRO-ED.

Carpenter, R. L. (1985). Mathematics instruction in resource rooms: Instruction time and teacher competence. **Learning Disability Quarterly,** 8, 95-100.

Cooney, J. B., & Swanson, H. L. (1990). Individual difference in memory for mathematical story problems: Memory span and problem perception. **Journal of Educational Psychology,** 82, 520-577.

Daniels, P.R. (1983). **Teaching the Gifted/Learning Disabled Child**. Austin, TX: PRO-ED.

Dinnel, D., Glover, J., & Ronning R. (1984). A provisional mathematical problem solving model. **Bulletin of the Psychonomics Society**, 22, 459-462.

Frank, K., & Wade, P. (1993). Disabled Student Services in Postsecondary Education: Who's Responsible for What? **Journal of College Student Development,** 34, 26-34.

Hammill, D. D. (1990). On defining learning disabilities: An emerging consensus. **Journal of Learning Disabilities**, 23, 74-83.

Hammill, D. D. and Bryant, B. R. (1992). **Detroit Test of Learning Aptitude - Adult**, Austin, Texas: Pro. Ed.

Hessler, G. L. (1984). **Use and Interpretation of the Woodcock-Johnson Psycho-Educational Battery**, Allen, TX: DLM Teaching Resources.

Heywood, Lawton, & Associates (Eds.). (1991b). Documentation on the need for academic adjustments. **Disability Accommodation Digest**, 1(3), 3. Columbus, OH: Association on Handicapped Student Service Programs in Postsecondary Education.

Hoover, J. J., (1993). Helping Parents Develop a Home-Based Study Skills Program. **Intervention in School and Clinic,** 28, 238-245.

Hutchinson, N. L. (1993). Effects of Cognitive Strategy Instruction on Algebra Problem Solving of Adolescents with Learning Disabilities. **Learning Disability Quarterly,** 16, 34-63.

Hutchinson, N. L. (1993). Effects of Cognitive Strategy Instruction on Algebra Problem Solving of Adolescents with Learning Disabilities. **Learning Disability Quarterly,** 16, 34-63.

Hughes, C.A. & Smith, J.O., (1990). Cognitive and Academic Performance of College Students with Learning Disabilities: A synthesis of the Literature. **Learning Disability Quarterly,** 13, 66-79.

Hutchinson N. L., (1993). Effects of Cognitive Strategy Instruction on Algebra Problems Solving of Adolescents With Learning Disabilities. **Learning Disability Quarterly**, 16, 34-63.

Jarrow, J. (1987). Integration of individuals with disabilities in higher education:A review of the literature, **Journal of Postsecondary Education & Disability**, 5(2) 38-57.

Jarrow, J. (1992). **The Impact of Scection 504 on Postsecondary Education**. Cloumbus, Ohio: Association on Higher Education And Disaabiliaty.

Kincaid, J. M. **Compliance Requirements of the ADA and Section 504.** Paper presented at the Association on Higher Education and Disability conference, Long Beach, CA.

Kirby, J.R., & Becker, L. D. (1988). Cognitive components of learning problems in arithmetic. **Remedial and Special Education,** 9(5), 7-16. Kochnower, J., Richardson, E., & DiBenedetto, B. (1983). A comparison of the phonic decoding ability of normal and learning disabled children. **Journal of Learning Disabilities**, 16, 348-351.

Leonard, F. C., (1991). Using Wechsler Date to Predict Success for Learning Disabled College Students. **Learning Disabilities Research & Practice,** 6, 17-24.

Malcolm, C. B., Polatajko, H.J., & Simmon, J. (1990). A descriptive study of adults with suspected learning disabilities. **Journal of Learning Disabilities**, 23, 518-520.

Mather, N. (1991). **An Instructional Guide to the Woodcock-Johnson Psycho-Educational Battery**. Bradon, VT: Clinical Psychology Publishing Co., Inc.

Maters, L. F., & Mori, A. A. (1986). **Teaching Secondary Students With Mild Learning and Behavior Problems: Methods, Materials, Strategies.** Rockville, MD:Aspen Systems.

Mayer, R. E. (1993). Understanding Individual Differences in Mathematical Problem Solving: Towards A Research Agenda. **Learning Disability Quarterly,** 16, 2-5.

Mercer C. D. (1991). Strategic Math Series: Part One-Examining Components of Effective Math Instruction. **Strategram, Strategies Intervention Model,** 1, 1-3.

Mercer, C. D., & Miller, S.P. (1991) Strategic Math Series: Part Two-Programming for Effective Teachiang of Basic Math Facts-. **Strategram, Strategies Intervention Model,** 4,1-8.

Meyers, M. J. (1987, November). LD students:Clarifications and recommendations. **Middle School Journal**, 27-30.

Miller, J. H., & Milam, C. P. (1987). Multiplication and division errors committed by learning disabled students. **Learning Disabilities Research**, 2, 119-122.

Montague, M., Bos, C., and Doucette, M.(1991). Affective, Cognitive, and Metacognitive Attributes of Eighth-Grade Mathematical Problem Solvers. **Learning Disabilities Research & Practice,** 6, 145-151.

Nolting, P. D. (1988). **Winning at Math:Your Guide to Learning Mathematics the Quick and Easy Way**. Bradenton, FL: Academic Success Press.

Nolting, P. D. (1989). **Strategy Cards for Higher Grades**. Bradenton, FL: Academic Success Press.

Nolting, P. D. (1989). **Math Study Skills Computer Evaluation Software**: Bradenton, FL: Academic Success Press.

Nolting, P. D. (1991). **Winning at Math: Your Guide to Learning Mathematics Through Effective Study Skills**. Bradenton, FL: Academic Success Press.

Nolting, P. D. (1991). **Math and the Learning Disabled Student: A Pratical Guide for Accommodations.** Bradenton, FL: Academic Success Press.

Norton, S. M. (1992). Postsecondary Learning Disabled students: Do Their Study Habits Differ From Those of Their Non-Learning Disabled Peers? **Community/Junior College Quarterly,** 16, 105-115.

Rourke, B. P., & Strang, J. D. (1983). Subtypes of reading and arithmetic disabilities:A neuropsychological analysis. In M.J. Rutter (Ed.), **Developmental Neuropsychiatry**, (pp473- 488). New York: Guiford Press.

Skrtic, T. M. (1980). Formal reasoning abilities of learning disabled adolescents: Implications for mathematics instruction. (Available from the Institute for Research in Learning Disabilities, University of Kansas, Lawrence, KS).

Sparks, R., Ganschow, L. and Javorsky, J. (1992). Diagnosing and Accommodating the Foreign Language Learning Difficulties of College Students with Learning Disabilities. **Learning Disabilities Research & Practice,** 7, 150-160.

Strang, J. D., & Rourke, B. P. (1985). Arithmetic disability subtypes: The neuropsychological significance of specific learning impairments in childhood. In B.K. Rourke (Ed.), **Essentials of subtypes analysis** (pp. 167-183). New York: Guiford Press.

Thornton, C. A., & Bley, N. S. (1982). Problem solving:Help in the right direction for learning disabled students. **Arithmetic Teacher**, 29(6), 26-27, 38-41.

Thornton, C. A., & Toohey, M. A. (1985). Basic math facts: Guidelines for teaching and learning. **Learning Disabilities Focus**, 1(1), 44-57.

Vogel, S. A. (1985a). Learning disabled college student: Identification, assessment, and outcomes. In D.D. Drake & C. K. Leong(Eds.), **Understanding learning disabilities: International and multidisciplinary views** (179-203). New York: Plenum Press.

Vogel, S. A., Hruby, P. J., & Adelman, P. B.(1993). Educational and Psychological Factors in Successful and Unsuccessful College Students with Learning Disabilities. **Learning Disabilities Research & Practice,** 8(1). 35-43.

Waldron, K. A., & Saphire, D. G. (1989). Perceptual and academic patterns of learning-disabled/gifted students. Manuscript submitted for publication.

Waldron, K. A., & Saphire, D. G. (1990). An analysis of WISC-R factors for gifted students with learning disabilities. **Journal of Learning Disabilities**, 23, 491-498.

Wechsler, D., (1981). Wechsler Adult Intelligence Scale-Revised. New York, NY: The Psychological Corporation.

White, W., Alley, G., Deshler, D. Schumakxer, J., Warner, M., & Clark, F. (1982). Are there learning disabilities after high school? **Exceptional Children**, 49, 273-274.

Woodcock, W., & Johnson, M. (1977). **Woodcock-Johnson Psycho-Educational Battery Achievement Tests**. Allen, TX: DLM.

Woodcock, W., & Johnson, M. (1989). **Woodcock-Johnson Psycho-Educational Battery Achievement Test - Reivsed**, Allen, TX: DLM Teaching Resourses.

Zawaiza R., & Gerber, M. (1993). Effects of Explicit Instruction on Math Word-Problems Solving by Community College Students With Learning Disabilities. **Learning Disabilities Quarterly**, 16, 64-79.

Zentall, S. S. (1990). Fact-retrieval automatization and math problem-solving: Learning disabled, attention disordered, and normal adolescents. **Journal of Educational Psychology.** 82, 856-865.

Zentall, S. S. (1975). Optimal stimulation as theoretical basis of hyperactivity. **American Journal of Thopsychiatry,** 45, 549-563.

Zentall S. S., & Ferkis, M. A. (1993). Mathematical Problem Solving For Children with ADHD, with and without Learning Disabilities. **Learning Disability Quarterly**, 16, 6-18.

AUTHOR BIOGRAPHICAL DATA

Dr. Paul Nolting

Over the past fifteen years Learning Specialist Dr. Paul Nolting has helped thousands of students improve their mathematics learning and obtain better grades. Dr. Nolting is an expert in assessing mathematics learning problems-from study skills to learning disabilities-and developing effective learning strategies and testing accommodations.

He is a national consultant and trainer of math study skills and of learning and testing accommodations for students with learning disablities. He has conducted national training grant workshops on learning disabilities for the Association on Higher Education and Disabilities and for a two year University of Wyoming 1991-1992 Training Grant of Basic Skills. Dr. Nolting has conducted numerous national conference workshops on learning disabilities and mathematics for the National Developmental Education Association and the National Council of Educational Opportunity Association. His text, **Math and the Learning Disabled Student: A Practical Guide for Accommodations** is used thoughout the United States and Canada as the definitive text for disabled student service providers.

Dr. Nolting holds an M.S. degree in Counseling and Human Systems from Florida State University and a Ph.D. degree in Counselor Education from the University of South Florida. He has completed an advanced graduate level course with Dr. Richard Woodcock on interpretations of the Woodcock-Johnson-Revised. His book, **Winning at Math: Your Guide to Learning Mathematics Through Successful Study Skills** was selected Book of the Year by the National Association of Independent Publishers. "The strength of the book is the way the writer leads a reluctant student through a course from choosing a teacher to preparing for the final examination," says **Mathematics Teacher**, a publication of the National Council of Teachers of Mathematics.

His two audio cassettes, **How to Reduce Test Anxiety** and **How to Ace Tests** were also winners of awards in the National Association of Independent Publishers' competition. "Dr. Nolting," says Publishers's Report, "is an innovative and outstanding educator and learning specialist."

A key speaker at numerous regional and national education conferences and conventions, Dr. Nolting has been widely acclaimed for his ability to communicate with faculty and students on the subject of improving mathematics learning.

Other Publications from Academic Success Press

Books

Winning at Math: Your Guide to Learning Mathematics Through Successful Study Skills This textbook teaches students how to improve their mathematics study skills. Some of the topics include ten steps to doing your math homework, ten steps to test taking, note-taking techniques, test anxiety reduction techniques and memory techniques. This book also has learning and testing suggestions for students with learning disabilities. **Winning at Math** can be used for individual study or as a class room text. $14.95 (1991)

Math and the Learning Disabled Student: A Piratical Guide for Accommodations This book is a one-stop resource book for the teacher, counselor, service provider, or other professionals who provide assistance to students with learning disabilities. Topics include understanding reasons for mathematics learning problems, providing mathematics learning accommodations, providing mathematics testing accommodation, and case studies. $14.95 (1991)

How to Develop Your Own Math Study Skills Workshop and Course This book is for service providers and instructors who want to teach student mathematics study skills through workshops or a course. This book provides examples of math study skills/test anxiety workshops, ideas for recruiting student, sample announcement flyers, math study skills course outline and a sample of a final exam. $12.95 (1990)

Audio Cassette Tape

How to Reduce Test Anxiety This audio cassette tape was developed to teach students how to relax and reduce their test anxiety. The tape features explanation of the causes of test anxiety, five quick and easy-to-learn relaxation techniques, and long-term relaxation procedures guaranteed to work. $9.95 (1989)

How To Ace Tests This audio cassette tape suggest test-taking techniques for all tests including mathematics. The tape features how to predict test question, prepare for tests, how to become test wise, and steps to better test taking. $9.95 (1989)

Winning at Math: Your Guide to Learning Mathematics Through Successful Study Skills The narrative of **Winning at Math** was developed especially for students with visual impairments and learning disabilities. $29.95 (1991)

Video Tape

Math Study Skills and Test-Taking Skills Video Tape Teaching mathematics study skills, test-taking skills, and analyzing test results can now be accomplished by watching a video tape. This video tape has two parts which are based on **Winning at Math**. Part I features preparation and test-taking skill and Part II features analyzing results and improving performance. This video can be used as part of a mathematics workshop or be individual viewed by students. Introductory price: $49.95 (1994)

Strategy Cards

Strategy Cards for High Grades (3" x 5") These strategy cards were developed for student with learning disabilities as learning and testing aids. The series of ten different cards capture invaluable rules for student study and test-taking. The cards are based on the **Winning at Math** text and make a handy supplement to any study skills course, math lab and can be uses as learning and testing accommodations. $4.95 (1988)

Strategy Cards for High Grades (8 1/2" x 11") These large strategy cards were developed for students with visual impairments or visual discrimination problems. The large strategy cards also make an excellent outline for information given in mathematics workshops and mathematics courses. The content of the large cards is the same as the small strategy cards. $9.95 (1989)

Computer Software

Winning at Math: Study Skills Computer Evaluation Software This is a computerized math study skills evaluation that can be personalized for each student. The program analyzes students mathematics study skills, identifies specific study problem areas, makes suggest for study improvement, and cites page reference in **Winning at Math** and suggest other materials to improve the students study skills. The program also suggests specific learning and testing accommodations for students with learning disabilities. Site license $74.95 (1989)

Kits

Winning at Math Study Skills Success Kit This success kit was developed to diagnose and provide treatment to college students with mathematics learning problems. Most students including students with learning disabilities will need help in several areas such as test anxiety reduction, study strategies, test-taking skills, learning accommodations, and test-taking accommodations to become successful in mathematics. The computer quickly identifies problem areas in student's study skills, and recommends specific solutions, citing other materials included in this kit. The kit includes **Winning at Math Study Skills Computer Evaluation Software, Winning at Math, How to Reduce Test Anxiety, How to Ace Test, and Strategy Cards for Higher Grades (3" x 5")**. Individually priced: $114.75 Success Kit price: $94.95

Mathematics Workshop/Course Development Kit Instructors and service providers sometimes have difficulty developing math workshops and math study skills courses. This kit already has the materials necessary to begin your own mathematics workshop or math study skills/test anxiety course. For example, the video tape can be use as part of the workshop and the strategy cards can be used as lecture notes. The **Winning at Math** teacher's manual would supplement the development of a math study skills course by having chapter reviews, additional topic discussions, and answers to the homework questions. The kit includes **Math Study Skills and Test-Taking Video Tape, How to Develop Your Own Math Study Skills Workshop and Course, Strategy Cards for High Grades (8 1/2" x 11"), Winning at Math,** and **Teacher's Manual For Winning at Math**. Introductory priced: $89.95

To order these materials send check/money order or purchase order to:

Academic Success Press, Inc.
P.O. BOX 25002, BOX 132,
Bradenton, Florida 34206

or call (800) 444-2524/FAX (813) 753-2882.

(Please include $4.00 first item / $1.50 each additional item, and $8.00 for each Kit for shipping.)